CONSTRUCTION MANUAL For Highway Construction

Developed by the Highway Subcommittee on Construction

Published by the

American Association of State Highway and Transportation Officials
444 N. Capitol Street, N.W., Suite 225
Washington, D.C. 20001

ISBN 1-56051-0021

HIGHWAY SUBCOMMITTEE ON CONSTRUCTION

1989

Chairman	Richard Trainor, Maryland
Vice Chairman	Robert L. Clevenger, Colorado
Secretary	Norman J. VanNess, FHWA

Alabama .... William J. Hartzog
Alaska .... Loren Rasmussen, David Cowee
Arizona .... August Hardt, Donald L. Cornelison
Arkansas .... Allan Holmes, Calvin Peevy
California .... Jerry Russell
Colorado .... Robert L. Clevenger, Glenn W. Fritts
Connecticut .... Arthur Hourihan, Walker W. Cox
Delaware .... R. C. McDowell
District of Columbia .... Gary Burch
Florida .... Robert E. Buser
Georgia .... Stanley Lord
Hawaii .... Edward Y. Asato
Idaho .... Richard Hedges
Illinois .... Edward J. Kehl
Indiana .... William J. Ritman
Iowa .... Harold Dowden
Kansas .... Dean M. Testa
Kentucky .... R. A. Walsburger, R. E. Back
Louisiana .... Earl Cryar
Maine .... John E. Hodgkins
Maryland .... James M. Welsh, Richard H. Trainor
Massachusetts .... William A. Billings
Michigan .... Gary Taylor
Minnesota .... Wayne F. Murphy
Mississippi .... Kenneth Warren
Missouri .... Gary Chullino
Montana .... William H. Larson
Nebraska .... R. J. Stutzman
Nevada .... Jack Montrose
New Hampshire .... Adolf B. Baer
New Jersey .... Albert B. Ari, John A. Walsh
New Mexico .... John Winton
New York .... Kenneth W. Shiatte, Michael Cuddy
North Carolina .... B. G. Jenkins, Jr., Len A. Sanderson
North Dakota .... Francis Ziegler

Ohio .... J. Keith Henry, George C. Young
Oklahoma .... D. C. Carman
Oregon .... Kenneth E. Husby
Pennsylvania .... Joseph Filippino
Puerto Rico .... Gilberto Carrero
Rhode Island .... Joseph R. Simeone
South Carolina .... Oren S. Fletcher, W. A. Keller, III
South Dakota .... Mike Durick
Tennessee .... Richard Omohundro, Johnie E. Davis
Texas .... Bobbie F. Templeton
U.S.D.O.T. (Federal Highway Administration) .... Roger Goughnour
U.S.D.O.T. (Federal Aviation Administration) .... Leonard Mudd
U.S.D.O.T. (Federal Highway Administration) .... Norman J. Van Ness
Utah .... Bert L. Taylor
Vermont .... Thomas K. Pierce
Virginia .... Claude D. Garver
Washington .... Del Vandehey
West Virginia .... Earl Scyoc
Wisconsin .... Heny O. Ellison
Wyoming .... Edward Crowe
Alberta .... N. Chorney
Guam .... Daniel A. Lizama
Hong Kong .... H. C. Beaton
Marianna Islands .... John C. Pangelinan
New Brunswick .... M. S. McInnis
Northwest Territories .... A. Gamble
Nova Scotia .... J. V. Gavin
Ontario .... Don Barnes
Saskatchewan .... M. J. Herasymuik
New Jersey Turnpike Authority .... R. Bruce Noel
Mass. Metro. Dist. Comm. .... John W. Wright
New York State Bridge Authority .... William Moreau
Forest Service .... John R. Holt

CONTENTS

INTRODUCTION TO THE MANUAL

This is the fourth edition of the AASHTO *Construction Manual for Highway Construction*.

This publication was drafted by a Task Force of the Highway Subcommittee on Construction, which has the assistance and cooperation of all of the States. This Manual is designed as a companion to the AASHTO *Guide Specifications for Highway Construction*.

It is the function of highway specifications to set forth those things required of the Contractor, the method of measurement, and the basis of payment, among other things.

On the other hand, the Construction Manual defines the responsibilities of, and instruction to, State highway or transportation department field personnel regarding the proper administration of Contract Provisions.

The publication is intended as a guide and is endorsed to the State highway or transportation departments for their use in preparing their own individual construction manuals. The individual State highway or transportation department may wish to include additional material, directions, or detail to meet their respective needs or requirements. The organizational structure described and the position titles used throughout the text are typical and intended only to simplify the task of writing. Each individual department will make those modifications necessary to accommodate their particular situation.

Terms used in the contract, specifications, and this manual are defined in Section 101 of the AASHTO *Guide Specifications for Highway Construction*.

DIVISION 100 — ADMINISTRATION, ORGANIZATION AND POLICY

Section 101 — Department Organization

The Department is organized for the purpose of planning, designing, constructing, and maintaining an adequate system of safe highways capable of meeting the traffic needs of the State.

It is organized under the overall supervision of a Chief Administrative Officer who is assisted by various Divisions, each under a responsible administrator.

The Division of Construction is charged with the responsibility for administration of contracts. It assumes responsibility at the time the contract has been awarded in accordance with procedures established by the Department. Its responsibility ceases when the project has been accepted by the Department under established policy.

The District Engineer is the operational representative of the Chief Administrative Officer in each of the various districts of the State. Among the District Engineer's assistants is one assigned to contract administration and designated as District Construction Engineer. The Division of Construction on behalf of the Chief Administrative Officer assists the District Engineer and the assistant in matters of policy, administration, and study of special problems associated with construction.

Operating under the general supervision of the District Construction Engineer is the Project Engineer who is the Department's representative on the project. Personnel will be assigned to the Project Engineer to assist in the staking and inspection of the work.

101.01 — Federal Highway Administration (FHWA)

On a highway project for which all or part of the funding is made available by the Federal Government, the terms of Federal participation are set up in an agreement between the Department and the FHWA. Such an agreement provides that the work is to be done in accordance with predetermined standards embodied in the plans and specifications, in other approved standard drawings, and in any special provisions required by the nature of the project.

Administration of the construction is a function of the Department and its engineers and inspectors. However, engineers from the FHWA will make inspections of these projects at times selected by them.

The relationship between the FHWA and the Department does not directly involve the Contractor. FHWA representatives inspect the project for the purpose of reviewing the Department's procedure requiring the project to be constructed in accordance with the commitments contained in the State-FHWA Agreement. The FHWA's representative is inspecting the State's performance and not the Contractor's. The FHWA has neither responsibility nor authority to deal directly with the Contractor.

Department employees should cooperate with FHWA representatives in their inspections. Their comments should be noted in the diary and matters that require action should be referred promptly to the District Engineer. Their attention should be called to necessary extra work and to any proposed changes. All major changes must be brought to the attention of the FHWA before any of the work is started, and concurrence of the FHWA before the effective date of beginning the work involved in the major change shall be obtained.

101.02 — Project Organization

Each contract is under the direct supervision of a Project Engineer. The Project Engineer is assigned to the project prior to the award of the contract. The assignment is documented by a letter from the District Engineer to the Project Engineer. The Contractor is notified by letter of the name and address of the Project Engineer assigned to the project.

The Project Engineer is responsible for assuring the project is constructed in accordance with the plans, specifications and special provisions, and control of inspection, and proper documentation.

The personnel assigned to assist the Project Engineer will have varying levels of training and experience. The Project Engineer will be responsible for the proper assignment of these personnel.

As the Department's representative, the Project Engineer will have frequent personal contacts with the Contractor, property owners, municipal officers, utility representatives, and the traveling public. The conduct of these associations must be of a character to reflect credit on the Project Engineer and the Department. Further discussion of these contacts will be developed later in this manual.

101.03 — Inter-Department Relations

Harmonious working relations among all employees of the Department are most important. An understanding of the functions and problems of other divisions, as well as the manner in which they fit into the overall organization, will improve the teamwork within the Department. Each employee has a responsibility to promote and foster good relations with fellow employees. An employee is expected to carry out the instructions of the supervisor. Each supervisor's conduct should earn full support and cooperation of all assigned employees. Each employee must know his/her responsibility and must have the authority to handle it.

A major factor in the promotion of good working relations is to keep your supervisor fully informed about all pertinent events that happen on work for which you are responsible. This principle applies equally at all levels of authority.

The Project Engineer should brief the assigned personnel on plans and schedules for the work immediately ahead.

101.04 — Department Equipment

Engineering and Vehicular Equipment

The Project Engineer has the responsibility of requiring personnel to learn and use correct practice for operation of the engineering and vehicular equipment and assuring that the equipment is properly protected and stored when it is not in use.

Section 102 — Bidding Requirements and Conditions

The Specifications establish the conditions under which bids are accepted by the Department and establish the Contractor's personal responsibility for knowledge of job conditions and familiarity with plans and specifications. The Project Engineer should be aware of the contents of this section even though there is no involvement in bidding procedures.

If an inspector or Project Engineer accompanies the contractor during a pre-bid review of the contract limits he must be aware that any statements made regarding conditions, work difficulty or any other comments not included in the Plans and Specifications, have the same effect as contract data and could influence the bid preparation.

Section 103 — Award and Execution of Contract

The Specifications outline the procedures and obligations involved in award of the contract to the successful bidder. The Project Engineer will know that these conditions have been met when an executed copy of the contract or an official notice that the Contractor may proceed with the work is received.

The Project Engineer should not allow work to start prior to receipt of the notice to proceed.

Section 104 — Scope of Work

The Specifications establish conditions under which alterations may be made and establish the Contractor's and the Department's obligations with respect to maintenance of traffic. Be alert to special problems involved in traffic handling and make recommendations to the District Engineer when changes appear justified.

Section 105 — Control of Work

105.01 — Responsibilities of the Project Engineer

Under the direction of the District Engineer and the District Construction Engineer, the Project Engineer shall have immediate charge of one or more construction projects. The Project Engineer will be held responsible for the activities and efficiency of subordinates and for the satisfactory administration and monitoring of the work.

When it is the Project Engineer's responsibility to stake the projects he must see that the Contractor is informed of the meaning of all stakes. If the Contractor is responsible for all or part of the stakeout, the Project Engineer shall monitor and assure thorough spot checks that lines, grades, and elevations are in conformance with plans and specifications. It is also the Project Engineer's responsibility to know and to document the fact that all materials used meet controlling specifications and that the finished project meets alignment, grade, quality, and quantity requirements of the contract. Authority to alter plans shall be limited to minor field changes. If changes in specifications or major alterations of plans appear necessary, conditions should be investigated and recommendations promptly submitted to the District Construction Engineer.

As the Department's representative, the Project Engineer must assure that the project meets all contract requirements. The Project Engineer should expect to obtain no more than what is specified or accept any less than the contract requirements. The Project Engineer shall in no way attempt to supervise work for the Contractor.

The Project Engineer will be held responsible for the accuracy of all notes and reporting procedures.

Each Project Engineer shall keep a diary for each contract in which matters of importance regarding the project shall be entered daily.

The Project Engineer is not usually expected to function as the sole representative of the State but needs various assistants, inspectors, and others to watch closely the different phases of work for proper compliance and for keeping records in order.

To have an efficient organization, the line of authority must be well-defined. Each employee should be delegated authority in line with administrative responsibilities. The Engineer must check to see that delegated duties are being properly discharged.

It is good policy to delegate the responsibility for inspection supervision on the project to one or more experienced inspectors. As Chief Inspectors, these persons must have the authority to direct and coordinate the activities of all inspection personnel and make day-to-day decisions involving engineering judgments of an immediate nature.

Although the Project Engineer does have a qualified inspector on the project to observe the work, frequent checks should be made of the inspection operations and reports, noting whether plans and specifications are being properly interpreted and applied. The Project Engineer should also be alert to any possible difficulties which could arise either in the construction or in the final function of the project and try to make correction before the situation grows worse.

105.02 — Duties and Authority of Inspectors

Competent inspection is one of the most important elements entering into any construction work.

The Project Engineer may serve as the chief inspector or may delegate this assignment to one or more assistants. If the Project Engineer serves as the chief inspector, a major portion of time shall be spent in the field on the project.

If faulty material or poor workmanship is incorporated into the construction, the work shall be removed or corrected. Proper inspection requires good judgment, diplomacy, common sense, and a thorough knowledge of the work, plans, and specifications.

It shall be the duty and authority of all inspectors on construction projects to enforce the specifications. If differences in interpretation arise with the Contractor, the matter shall be decided by the Project Engineer.

It is not good practice for the Project Engineer to use the same inspectors on successive jobs with the same Contractor. Certain personal relations and precedents can be established which may not be in the public interest.

The Inspector shall check everything carefully to see that it conforms with the specifications. The Inspector shall have the authority to reject materials and to suspend any work which does not conform with the specifications.

When the methods of procedure in any operation are prescribed in the specifications, they shall be rigidly enforced. The Inspector shall always bear in mind that the management of the work is the Contractor's business; however, if any methods are employed which the Inspector has reason to believe will impair the quality of the finished job, the Contractor shall be advised accordingly and the Project Engineer notified immediately. The inspector shall in no way attempt to supervise work for the Contractor.

The Inspector shall not be authorized to revoke, alter, enlarge, relax, or release any requirements of the plans or specifications, nor to approve or accept any portion of the work or to issue any instructions contrary to the plans or specifications. If conditions arise which seem to render it impractical to enforce the specifications, the Inspector should contact the Project Engineer at once.

All Inspectors on construction projects shall work under and report directly to the Project Engineer or to a delegated Chief Inspector.

A set of plans and a copy of the specifications shall always be available for reference.

The Inspectors shall keep a diary in which matters of importance shall be entered daily.

It shall be the duty of all Inspectors to study the plans and specifications of the job to which they have been assigned and they must be fully conversant with all the details of the work to be done. If anything is found which is not fully understood, the Inspector shall consult with the Project Engineer.

105.03 — Preconstruction Conference

A. As soon as possible after the project has been officially awarded, the Project Engineer shall arrange a conference with the Contractor and interested parties for the purpose of reviewing construction details, proposed schedules, etc. Prior to this meeting, the Project Engineer should have studied the Plans and Special Provisions and made a field inspection of the project in order to be well-informed as to the requirements and existing conditions.

B. Among those who should be invited to the conference are the following:

1. A representative of the District Office
2. Principal Assistants of the Project Engineer
3. The Contractor and principal personnel
4. The Representatives of involved utilities
5. Municipal and/or County Engineer, if involved
6. FHWA representative, if involved
7. Any other interested parties involved in the contract

C. Among the subjects to be discussed as applicable on the project are the following:

1. Project Engineer's general overview of the plan concept, specifications, traffic control procedures, and special problem areas.
2. Contractor's proposed operating schedules, computation of work day charges, execution of working day statements, time schedule, and completion date requirements, etc.
3. Work to be sublet, stipulations to be included in the subcontract agreements, Engineer-Contractor relations and responsibility towards subcontractors, authorized representatives, etc.
4. Labor Provisions, necessary posters, Engineer's inspection and investigating procedures with regard to labor requirements. The Contractor and the contracting authority shall be thoroughly familiar with the applicable EEO specifications and both will diligently pursue the requirements as intended.
5. Legal relations and responsibilities; cooperation with utility owners, the public and other Contractors; licenses and permits in connection with execution of the work, local ordinances, etc.
6. Special requirements and unusual conditions, conflicts and problems anticipated, clarification of construction details and specification requirements, procedures for assessment of time, etc.
7. Inspection procedures, notification to the Engineer of material orders, furnishing samples, and the time and place of testing and accepting materials, locating and equipping the field laboratories, storage and use of materials, etc.
8. Thorough discussion of plan concept for traffic control, maintenance thereof, and responsibilities of parties involved.
9. Haul road requirements; location and scheduling of bypass construction, crossroad closures and access facilities; general responsibilities with regard to traffic and public convenience, etc.

10. Employee and public safety, sanitary provisions, etc.
11. Delegation of authority by the Contractor and the Engineer, lines of communication, equipment, and personnel, etc.
12. A list of suppliers should be furnished the Engineer indicating where the Contractor proposes to obtain all materials for the project.

D. The Project Engineer is also responsible for the conference agenda, for conducting the discussions, and for making a written record of the conference discussions. The written record is prepared in letter form to the District Engineer with copies to all participants and to the project file.

E. The preconstruction conference, if properly conducted, can be of material aid in getting the project properly started. Participants should come prepared to make worthwhile contributions to the conference and improvement of general relations. As moderator, the Project Engineer should attempt to keep within the scheduled agenda once the conference has begun and discourage any extraneous or digressive commentary without becoming rude.

Section 106 — Control of Material

The Specifications establish the respective obligations of Contractor and Department concerning materials to be used.

Section 107 — Legal Relations and Responsibility to Public

107.01 — Integrity

Absolute integrity on the part of all Department personnel is essential if public confidence in the Department is to be maintained. Integrity embraces everything that touches a person's ability to do his/her job: resourcefulness, decisiveness, adaptability, stability, forcefulness and moral courage.

Department personnel should not engage in outside work unless the matter is cleared previously by the District Engineer. Project personnel are prohibited from doing engineering work for, and receiving compensation from, the Contractor.

No one in the Department whose job involves negotiating, approving or administering any contract or transaction on behalf of the Department shall have any financial or personal interest, direct or indirect, in the case.

If any Department employee has any interest in real property to be acquired for highway purposes, full documentation of the facts and circumstances of the interest shall be made. The employee shall not participate in acquisition of the property as agent of the State.

No employee shall use Department equipment for personal business or shall seek to acquire Department property which is offered for sale without express written approval of the Chief Administrative Officer.

The solicitation or acceptance of a cash loan by a Department employee from any Contractor doing business with the Department, or from any of the Contractor's representatives, is another example of serious conflict of interest.

Any case of dishonesty or conflict of interest will result in the immediate dismissal of the person or persons involved.

107.02 – Hatch Act

By virtue of the use of Federal-Aid funds for highway construction by the Department, employees are subject to provisions of the Hatch Act, a Federal Law concerning political activity.

TITLE 5, UNITED STATES CODE

CHAPTER 15.—POLITICAL ACTIVITY OF CERTAIN STATE AND LOCAL EMPLOYEES

Section 1501. Definitions. For the purpose of this chapter [§§ 1501-1508 of this title]—

(1) "State" means a State or territory or possession of the United States;

(2) "State or local agency" means the executive branch of a State, municipality, or other political subdivision of a State, or an agency or department thereof; and

(3) "Federal agency" means an executive agency or other agency of the United States, but does not include a member bank of the Federal Reserve System;

(4) "State or local officer or employee" means any individual employed by a State or local agency whose principal employment is in connection with an activity which is financed in whole or in part by loans or grants made by the United States or Federal agency, but does not include—

(A) an individual who exercises no functions in connection with that activity; or

(B) an individual employed by an educational or research institution, establishment, agency, or system which is supported in whole or in part by a State or political subdivision thereof, or by a recognized religious, philanthropic, or cultural organization; and

Pub.L. 89-554, Sept. 6, 1966, 80 Stat. 403; Pub.L. 93-443, Title IV, § 401(c), Oct. 15, 1974, 88 Stat. 1290.

Section 1502. Influencing elections—Taking part in political campaigns—Prohibitions—Exceptions.— (a) A State or local officer or employee may not—

(1) use his official authority or influence for the purpose of interfering with or affecting the result of an election or a nomination for office;

(2) directly or indirectly coerce, attempt to coerce, command, or advise a State or local officer or employee to pay, lend, or contribute anything of value to a party, committee organization, agency, or person for political purposes; or

(3) be a candidate for elective office.

(b) A State or local officer or employee retains the right to vote as he chooses and to express his opinions on political subjects and candidates.

(c) Subsection (a)(3) of this section does not apply to—

(1) the Governor or Lieutenant Governor of a State or an individual authorized by law to act as Governor;

(2) the mayor of a city;

(3) a duly elected head of an executive department of a State or municipality who is not classified under a State or municipal merit or civil-service system; or

(4) an individual holding elective office. (Sept. 6, 1966, Pub.L. 89-554, 80 Stat. 404; Pub.L. 93-443, Title IV, § 401(a), October 15, 1974, 88 Stat. 1290.

Section 1053. Nonpartisan candidacies permitted.—Section 1502.(a)(3) of this title does not prohibit any State or local officer or employee from being a candidate in any election if none of the candidates is to be nominated or elected at that election as representing a party any of whose candidates for presidential elector received votes in the last preceding election at which presidential electors were selected. Pub.L. 89-554, Sept. 6, 1966, 80 Stat. 404; Pub.L. 93-443, Title IV, § 401(b)(1), Oct. 15, 1974, 88 Stat. 1290.

107.03 — Public Relations

A good Department-contractor relationship can be maintained if these suggestions are considered:

a. Treat the Contractor fairly and impartially
b. Study the Contractor's viewpoint and be friendly but impersonal. Do not put yourself under any obligations.
c. Do not discuss with outsiders the Contractor's methods of handling work. Make your suggestions to the Contractor only.
d. Be ready to advise the Contractor when requested, but do not make snap decisions.
e. Issue all orders only to the Contractor or authorized representative.
f. Write and retain copies of specific orders given.
g. Discuss the Contractor's schedule. Coordinate your staking and inspection with the schedule.
h. Do not accept gratuities from the Contractor; do not be threatened or intimidated. Notify your supervisor of any trouble.

Courtesy and consideration are necessary in all contacts with the public as the entire Department is judged by the actions of its employees. Although an employee may not obligate the Department to any course of action or any expense without due authorization, it is necessary to always be as tactful and helpful as conditions permit. The media of communication with the public are many: The newspaper, the radio, television, service clubs, Chambers of Commerce, city and county officials, and direct contact with individuals.

If conditions are observed that might develop into public controversy and misunderstandings, this information should be transmitted through channels so that early news releases can inform the public of the facts. It is important that information given to the public not be slanted or evasive.

In the case of contacts with newspapers, radio, or television, the Project Engineer should furnish information only on matters for which he/she has personal responsibility and in which he/she is well informed. Questions concerning policy or programs should be referred to the District Engineer for consideration.

In dealing with residents along the highway, the Project Engineer must endeavor to maintain friendly relations. Frequently requests will be made on which the Project Engineer will not have the authority to act. In such cases the Project Engineer must make every effort not to offend people making these requests. The Project Engineer should not refer the residents to his/her supervisor. However, the residents should be advised that the matter will be discussed with the supervisor. Subsequently, the Project Engineer should personally inform interested residents of the results of the discussion.

107.04 — Labor Regulations

As part of the contract, it is required that all regulations with regard to labor be properly followed. Requirements of the Contractor and the State regarding the contract labor provisions are well covered in the proposal and the referenced regulations. A working knowledge of these regulations is expected of the State personnel involved to assure compliance by the Contractor.

Also included in the contract provisions are various requirements that relate to equal employment opportunity.

The following subsection will provide further explanation of regulations involving the above.

107.05 — Required Notices and Posters

The proposals for highway construction contracts state that certain information must be displayed in a conspicuous place on the project so that interested persons may be readily aware of their contents. This information is presented on the following:

A. Fraud Poster. A "Fraud Poster" required by Section 1020, Title 18, United States Code, must be displayed during the course of the work. It points out the consequences of impropriety on the part of any employee working on the project. For the purpose of advising people where they can report a violation, should any arise, the addresses of the State Highway or Transportation Department and the Federal Highway Administration are shown on the form.

B. Wage Rate Information. The schedule of wage rates, as shown in the proposal and as subsequently modified or amended, shall be posted where all the workers can view it. If more than one wage area is listed, it would be well to clearly define the wage area applicable to the project. For the purpose of advising the employees where they can report a wage violation, should any arise, the name and address of the State Highway or Transportation Department representative is shown on the form.

C. Equal Opportunity Posters. The poster titled "Equal Employment Opportunity is the Law" shall be posted on the project where all the workers can view it. Posters or Notices relative to the Company EEO Officer, etc., should also be posted on the job site.

107.06 — Enforcement of Labor Provisions

In order to properly fulfill the contract, the Contractor must conform to the labor provisions included therein. It is thus the Project Engineer's responsibility to be certain that the requirements regarding labor are properly carried out. Checks and documentation of the labor regulations take a variety of forms as follows:

A. Posters. The required posters and wage rates must be posted on the project when work is in progress.

B. Interviews. The Project Engineer and his/her staff should conduct random on-the-job interviews with several employees of the Contractor at appropriate intervals to determine the actual wages being paid and whether the employee is properly classified in the work being done. This information is reported on proper forms with copies going to the District or Central Offices. Any wage discrepancies should be brought to the attention of the Contractor without delay.

C. Statement of Compliance. The Contractor is required to submit a weekly Statement of Compliance on all Federal-Aid Highway contracts containing a wage rate decision. This affidavit relates to Antikickback regulations and is required throughout the course of the work.

D. Weekly Certified Payrolls. The Contractor must, on all Federal-Aid Highway contracts containing a wage rate decision, furnish each week a statement to the Project Engineer with respect to the wages paid each of its employees (including apprentices and trainees) engaged on the work during the preceding weekly payroll period. The Statement of Compliance as described above must accompany the weekly statement.

Prior to forwarding of the payroll and Statement of Compliance to either the District Office or the Central Office, the Project Engineer or his staff should examine the payroll(s) for non-conformance with the wage provisions of the contract. A checklist to use in this determination of compliance is as follows:

1. Entries to be checked on every payroll:
   a. Work classification, title, code number for each employee;
   b. Hourly wage rates for each employee, including fringe benefits, if applicable;
   c. Are daily and weekly total hours shown?
   d. Is the certification signed?
   e. Are deductions itemized?
   f. Are all deductions approved?

Approved deductions are shown under Title 29, Subtitle A, Part 3.5 of the Federal Regulations.

2. At least the first two payrolls submitted by each Contractor or Subcontractor on each project should be fully checked with respect to:
   a. Arithmetical accuracy. In cases where the employer utilizes a computer printout type of payroll, it will not be necessary to check the extensions;
   b. Overtime computations;
   c. Wage rates verified with those shown in the applicable contract;
   d. Employee's full name, address, and social security number being shown on the payroll where the name first appears.

If full compliance is shown on the initial payrolls, the steps listed above can be performed on a sampling basis on subsequent payrolls, at the Project Engineer's discretion.

The project interview forms (refer to No. B above) must be cross-checked against the rates shown on the weekly payrolls.

E. Corrections. If it is found that the payroll has discrepancies and clerical errors, such errors should be called to the attention of the Contractor(s), and corrections should be made promptly. The original submitted payroll should not, under any circumstance, be returned to the Contractor. Corrections are to be made by supplemental payrolls prepared and submitted in the same manner as the original. It is necessary that a completely revised payroll transcript be submitted. The payrolls are to be submitted to the Project Engineer within 7 days following the end of the pay period. The payroll period usually ends on Friday or Saturday, and the Contractor should thus submit payrolls to the Project Engineer not later than the following Friday for checking. The payroll(s) should be transmitted as soon as possible after adequate checking has been done.

Delay in prompt submission of the required payrolls by the Contractor or Subcontractor may cause delay in the payment of monthly estimates to the Contractor.

F. Employment Records. Terms of the contract require that the Contractor's employment records be available during the progress of the work. The Project Engineer should make spotchecks of these records to establish:

1. Whether the wages being paid to the laborers and mechanics are not less than required in the contract;
2. If the work being performed by any specific class of employees, including helpers or apprentices, conforms to the classification set forth in the contract for the wage they are being paid;
3. Whether the age classification is correct if there appears to be a disproportionate number of lower paid help (laborers, helpers, apprentices) indicating avoidance of the minimum wage rules.

G. Violations. The Project Engineer is to investigate any complaints of violation of the Labor Standards. A report of each investigation and action taken is to be prepared and submitted to the District Engineer, and/or appropriate person in the Central Office. The provisions indicate the disposition in case there is substantial evidence that the violations with regard to payrolls are willful or if restitution by the Contractor or Subhcontractor is not made. Deliberate violation of the labor requirements regarding wages is a serious matter and cannot be tolerated.

H. Determining Wage Classification. In checking payroll and hearing complaints with regard to wage difficulties, the Project Engineer will be confronted with the problem of determining the actual classification in which an employee is working. Seldom is the solution clear-cut, nor has the Project Engineer authority to determine definitely the classification in which the Contractor's employee fits. The Project Engineer does not have information as to the duties of a Carpenter's helper as against those of a Carpenter, as an example, so that the final analysis in case of wage dispute must be resolved by others.

When a wage dispute arises, the Project Engineer should obtain and assemble all the information available, and consult the District Office, and/or the Central Office concerning the problem.

I. Equal Opportunity Regulations. Executive Orders 11141, 11246, and 11375 prohibit discrimination in employment because of race, color, religion, sex, age, or national origin.

More specifically, the contract provisions address themselves to the FHPM 6-4-1-2. The requirements set forth in this Order constitute the specific affirmative action requirements for project activities under the contract and supplement the equal employment opportunity requirements set forth in the required Contract Provisions (Form PR-1273 or 1316 as appropriate).

In some areas, the Contract Provisions may, in lieu of FHPM 6-4-1-2, carry specific contract provisions that pertain to hometown plans. These type of provisions will contain goals and timetables of minority manpower utilization required of the Contractor and Subcontractors.

J. Training. Depending on certain circumstances, the Contract Provisions may contain the requirements of the FHPM 6-4-1-2. The purpose of this Order is to provide on-the-job training, with emphasis directed toward disadvantaged and minority groups, in the skilled trades used by the highway industry.

There are two types of training programs that may be used to fulfill the training obligation. One type is the Union Apprenticeship Program which has been approved by the Department of Labor, while the second type is referred to as an On-the-Job Training Program. The On-the-Job Training Program is designed by the Contractor and approved by the Department.

107.07 — Safety Precautions

A. Personnel. Many hazards exist on a construction project which require all personnel to be constantly alert if injury is to be avoided. Appropriate safety requirements shall be an emphasis area on all projects. Some special precautions which can be taken to improve safety in hazardous situations are:

1. Wear hard hats around overhead operations such as bridges or batch plants.
2. Wear safety vests while working in traffic.
3. Post signs and flaggers when work must be done on existing highways.
4. Provide adequate illumination for night work.
5. Provide for a shoring of trenches that may be a hazard to workers.

Option:

6. Become familiar with basic requirements of the safety and health regulations for construction as established by the U.S. Department of Labor as they apply to State highway activities. This information can be found in the "Informational Guide on Occupational Safety on Highway Construction Projects."

B. Traveling Public. The traveling public should be protected from danger due to construction operations. Adequate barricades and signs should be placed where they are most effective. This should be done in accordance with Part VI of the Manual on Uniform Traffic Control Devices. If needed, flaggers should be provided. There should be no doubt when roads are completely closed and where detours are located. When signs and barricades have served their purpose, they should be removed. (See Section 618, Traffic Control)

Section 108 — Prosecution and Progress

The Specifications establish the Contractor's responsibility to furnish adequate forces for meeting specified project schedules as well as quality.

Unsatisfactory progress should be directed to the Contractor's attention in writing by the Project Engineer. If the Contractor fails to act in keeping with the intent of this section the matter should be directed immediately to the attention of the District Construction Engineer.

All time assessments must be supported by adequate diary information or other supporting documentation. Review of project history often must rely heavily on diary entries.

Section 109 — Measurement and Payment

109.01 — Force Account

All extra work, to be performed by the Contractor on a Force Account basis, must be approved before starting.

The approval request shall include a brief but complete description of the work to be performed, a definite location by station numbers and an estimate of cost, broken down to show costs for labor, materials, and equipment. The estimate should be as accurate as possible based on available information.

A statement shall set forth the reasons for the extra work. When necessary, a detailed explanation should be included in the letter of transmittal.

Force account work will be paid for in accordance with the contract provisions. The Project Engineer will keep accurate daily records of the work as it is done. This record will be submitted in duplicate to the Contractor's representative each day for signature. The Project Engineer and the Contractor will each receive one signed copy.

All bills for materials used must be properly supported by copies of the invoices for the materials received on the job. Freight costs are considered to be part of the cost of materials used on force account work.

All bills for labor on force account work shall show the name of the employee, the job classification, the rate of pay actually paid by the Contractor, the dates on which the work was performed, and the number of hours worked by each employee each day on force account.

All bills for equipment used on force account shall show the number of hours each piece of equipment worked each day, the dates on which the work was performed, the rate of pay for each piece of equipment, and sufficient description of the equipment so that the rate of pay can be checked. It is suggested that the bill show the manufacturer's name, model number, and size rating (weight, horsepower, or cubic yards) for the equipment. The rate of pay for equipment shall conform to the rates previously established by the Department or agreed upon in writing prior to start of the work.

At the conclusion of the work, the entire force account charge shall be summarized on forms provided for that purpose. The Contractor shall furnish invoices, payrolls, freight bills, etc., to support the charges.

109.02 — Project Diaries

The Project Engineer is required to keep a daily diary on each project or combination of projects. Each Inspector assigned responsibility for a major operations shall prepare a separate daily diary. The Project Engineer's diary need not repeat the Inspector's detailed entries but may contain only general information about these operations.

Make all entries directly into a bound book. Daily entries must be signed by the person making such entries.

The diary should contain a day-by-day record of all significant items relating to the project. Since it may become important evidence in the decision of claims or the establishment of responsibilities or liabilities, it is essential that the notes be complete.

A partial list of items to be noted in a diary are:

a. Weather
b. Orders given the Contractor
c. Important discussions with Contractor or representative, such as disputes
d. Official visitors and inspections
e. Work or materials rejected and reasons
f. Time of shutting down or resuming work and explanations
g. Work done by Contractor's forces during day
h. General purpose of work
i. Account of any time spent by Contractor's personnel or equipment on disputable items or work
j. Length and cause of any delay
k. Arrival and departure of major equipment and work crews
l. Record of telegrams and telephone calls
m. Unusual conditions, if any, such as high water, bridge failures, slides, etc.
n. Progress of stakeouts and surveys made
o. When appropriate, daily surveillance of traffic control as to effectiveness

At the completion of the project all diaries shall be filed as part of the permanent record.

109.03 — Field Notes

Field notes are the written record of pertinent information, measurements, and observations on the project. They should be kept according to uniform practices and conform as a minimum to the following general requirements.

1. **Neatness.** Use a sharp pencil of at least 3-H hardness. Avoid crowding. Keep the book as clean as possible.
2. **Legibility.** Use standard symbols and abbreviations to keep notes compact. Use plain lettering to avoid confusion.
3. **Clarity.** Plan work ahead so that data can be clearly indicated. Do not make ambiguous statements. Line up descriptions and make sketches for clarity. Record data in a consistent way. Assume that the person who will use your notes has no familiarity with the work.
4. **Completeness.** Show all pertinent measurements and observations. Use a degree of accuracy consistent with the operation. If in doubt about the need for the data, record it. Review data before leaving the field.

All entries must include:

a. The date
b. Weather conditions
c. Names of all persons in the party
d. Signature of person making the entries

The title page must be completed as the book or project is started. The book must be adequately indexed and cross-referenced as to contents.

5. **Permanence.** All entries should be made directly into bound books. At completion of the project they should be filed as part of the permanent record in keeping with Department policy.
6. **Honesty.** Record exactly what was done at the time it was done rather than depending on memory at a later time. Never erase. If an item is incorrectly entered, draw a line through the item and insert the corrected value immediately above. When it is necessary to add data to notes previously prepared, the additional item should be dated and initialed. Always enter notes directly into the record.
7. **Self-Checking.** Notes should be so kept that the work can be checked without returning to the field. Any person familiar with the project should be able to verify the accuracy of the work from the information contained in the notes. Use positive controls.
8. **Pride.** Strive to turn in a set of notes of which you can be proud. These records often create an impression on others concerning your ability and integrity.

109.04 — Other Records

The Project Engineer will require documents to substantiate acceptance of materials and work items for both quality and quantity.

Basis of acceptance for quality, the method of measurement and the basis of payment are established by specifications.

Documentary evidence on which to base payment shall be in accord with policies established by the Department. These must provide evidence of positive control.

A tabulation confining the policies is desirable. Such a tabulation should show as a minimum the Standard Contract Item, how pay quantities are determined, and what documentation is necessary. A typical example is shown herein.

Other records which are the Project Engineer's responsibility include recurring reports such as:

a. Progress reports
b. Payment estimates
c. Personal reports

Payment estimates must be supported by documentary evidence that work items allowed have actually been performed. Such evidence may take the form of scale tickets, diary entries, material receipts, etc.

For example, earthwork quantities might be supported by load count entries in the Inspector's diary or by diary entries which show limiting stations of completed balances.

Weight or volume tickets provide a sound basis on which to allow payment for items measured in this manner. Examples of some documentation requirements for pay items are shown on succeeding pages.

The Project Engineer must be sure that all items allowed for payment have been properly inspected and approved.

Whenever payment is allowed based on an estimate, the basis of the estimate must be well defined to assure that over payment to the contractor is avoided.

Avoid overpayment. Where work remains to be done, such as filling the holes in concrete, the amount allowed for the appropriate pay item should be reduced by the amount estimated as necessary to cover the remaining work. Full allowance should be made for completed items.

All pay estimates must be signed by the Project Engineer as evidence of his approval for payment of the quantities shown.

Computers are increasingly being used for the maintenance of these and other project records and for estimate preparation. In such cases, protected backup files and computer security is extremely important.

109.05 — Retention of Records

At completion of a project all data shall be placed in the district files for possible future reference. Items to be retained shall include:

1. A complete set of final (as built) plans. This will include location sketches, typical sections, tabulation of quantities, plan and profile, bridge plans, special layouts, and a complete set of original contract plans.
2. Project contract, construction correspondence, and payment estimates.
3. All Project Engineer progress reports.
4. Weight or volume delivery tickets for materials. (Such as aggregate base, asphaltic concrete, etc.)
5. Project Engineer and Inspector Diaries.
6. Construction field notebooks such as alignment, original and final cross sections, structure staking, etc.
7. Any other records pertaining to pay quantities.
8. All material test reports.

Five years after completion of the project, the routine correspondence and all except the final estimate and final change order may be discarded provided there is no conflict with Federal regulations. On Federal-Aid projects no records should be discarded, less than three years from the date of the final payment of Federal funds. See Part 17, Title 23, Code of Federal Regulations. At the same time the records mentioned above are discarded, all other records noted under items 3, 4, 7, and 8 above may also be discarded.

Important correspondence and all other records shall be retained indefinitely.

Standard Contract Items	Unit	Documentation	Determination of Pay Quantities
Clearing Swamp Clearing Grubbing Deep Grubbing	Acres or sta.	Stationing—Dimension and computations recorded in Miscellaneous Removal Book-Inspector's Daily Report or Inspector's Daily Record of Contract Removal Items signed and dated.	Measured in horizontal area bounded by lines 10 feet outside the line of tree trunks of the trees cut or stumps grubbed or measured along the road centerline in stations of 100 feet. Stay within the right-of-way. When clearing or grubbing occurs only on one side of the road, it is measured on the basis of one-half the distance along the road centerline.
Earth Excavation Subgrade Undercutting Earth Stripping	Cubic yards	Field books containing the cross section notes, the plotted cross sections, and any notes and computations made during the construction operation. All notes to be checked, signed, and dated.	Measured in cyds. of volume in its original position, computed by the method of average end areas, except where this method is not feasible. In the case of hydraulic fills, or other cases where it is not feasible to

Standard Contract Items	Unit	Documentation	Determination of Pay Quantities
			measure excavation in its original position, the volume will be measured as compacted embankment computed on the basis of the required section.
Precast Concrete Piles Furnished	Lin. feet	Foundation Piling Record Form signed by Resident Engineer.	Will be ordered pile length in lineal feet in accordance with the lengths determined by the Engineer to be furnished.
Pipe Culverts (Concrete or Vitrified Clay)	Lin. feet	Location, number of pieces, commercial length reported on Inspector's Daily Report. Report of shipment from tested stock of size and class need, and sealers for joints. Notice of Acceptance by Office of Testing and Research.	Measured by length in lineal feet, which shall be determined by multiplying the number of units in the structure by their commercial length.

Section 110 — Archeological and Paleontological Salvage

It is in the public interest to preserve for public use historical and prehistoric objects such as Indian ruins, sites, buildings, artifacts, fossils, or other objects of antiquity that may have significance from a historical or scientific standpoint. On a construction project when it appears that significant historic, or prehistoric, objects have been or are about to be encountered, the Project Engineer should immediately take steps to preserve them and should notify the District Engineer.

The District Engineer, in turn, should advise the appropriate authorities within the State of the facts and permit them to inspect the site for the purpose of determining the advisability of salvaging the objects.

Section 150 — Field Work and Staking

150.01 — General

A very important duty of field forces on construction work is proper and accurate staking of the project. Time is important in all contracts; likewise, the staking and handling of related field work is important because the progress of Contractor's forces is dependent upon adequate and proper staking. The contracting authority may require the Contractor to provide the necessary construction staking. This may be incidental or bid as lump sum in the contract.

Cooperation with Contractor. When a project is assigned, the Contractor should be contacted immediately to learn the plan of procedure. This should enable staking to start well in advance of his operations. If the contract provides for the Contractor to furnish the stakes, arrangements should be made for immediate delivery.

It is essential that the Contractor's personnel understand the method used in staking and the markings on the stakes. The work in the field should be discussed with the Contractor's representative, carefully explaining the staking. If there is any doubt of a full understanding, a written explanation should be given the Contractor. A written record leaves little doubt and may prove to be a valuable record in the event the work is not built to proper lines and grades as staked in the field.

Checking Plans. Before staking is started, a preliminary check of the plans for the project should be made in the field for apparent ommissions or corrections. A careful check should be made for proper location of drainage structures. Observe drainage areas to make certain structures are adequate.

Entrance locations should be checked and notations made of any additional requirements necessary to serve new properties which are not shown on the plans but which were in existence at the time the right-of-way was secured. No additions, deletions, or revisions should be made to any entrances to condemned property or to controlled access right-of-way without the approval of the District Construction Engineer.

150.02 — Organizing Staking Parties

A. General. The construction staking party has four main purposes:

1. Reestablish sufficient horizontal and vertical control points to assure convenient and accurate staking during construction and verify the existing horizontal and vertical control for accuracy.
2. Provide the Contractor with sufficient stakes of proper alignment and grade so the job can be constructed according to plans.
3. Take necessary measurements so that quantities can be accurately computed for payment.
4. Perpetuate the horizontal and vertical control points after completion of the project.

The Project Engineer should organize the staking party according to conditions and available personnel rather than keeping a set number on each party regardless of the type of work.

B. Field Notebooks. All field notes should be recorded in standard field notebooks. Loose-leaf books are not to be used for permanent records or for temporary notes to be copied later into the permanent record. Each notebook should be identified on the front cover as to subject (alignment, cross sections, etc.), project, route, and county. This information should be repeated on the first page of the book along with the Project Engineer's name and address to assure the book's return in case of loss. The next page should carry an index of the book's contents. The date, weather, and party personnel must be recorded in the upper right-hand corner of the first right-hand page of each day's work.

On larger projects, separate notebooks should be used for each operation; such as, alignment, cross sections, structures, etc. On smaller projects, some of the separate operations may be combined in a single notebook. Quite often on supplementary construction, the preconstruction cross sections, bench levels, slope stakes, and final cross sections may be combined in one field notebook. This is not only economical but eliminates filing notebooks at the District Office with only a few pages used.

Notes should be legible and recorded in such a manner that other engineers may readily interpret any part. Erasures should not be made in field books. If it becomes necessary to revise or delete any information, it should be crossed through and an explanation made. Good complete notebooks are valuable project records and should always be kept clean and dry, and stored in a fireproof safe.

C. Survey Equipment. When organizing the staking parties, considerable thought should be given to the survey equipment to be supplied to the parties. The equipment may consist of electronic distance and angular measuring instruments, transits, theodulites and levels. The most important equipment is the transit and level; particularly the transit, for in an emergency it can also be used for leveling work. Three types of tapes are used: the 100-foot steel tape or chain, the 50-foot steel tape generally graduated in inches for use on structure work, and the 50-foot metallic tape graduated in tenths for use in measuring earthwork. Other small tools such as range poles, level rods, plumb bobs, sledge, axe, brush knife, etc., depend on the particular job at hand. The available equipment should be checked to make sure it is accurate and in good condition.

150.03 — Establishing Centerline and Right-of-Way

A. *Office Work.* Before staking is started, plans should be checked and notebooks prepared. A set of plans should be available for the use of the staking party at all times. Notation of errors which are found in staking should be made on the field plans as well as in the notebooks. All curve and alignment data shown on the plans should be checked before staking starts.

B. *Alignment Book.* A well-prepared alignment book will permit the party to proceed without delays. Leave plenty of space in the notebook as inadequate information is caused by crowded notes. An accepted method for recording information in the notebook is shown in Figure 150-1.

The station number of the beginning of the project should be obtained from the plans. This station number should be inserted on a line near the bottom of the second or third left-hand page of the field book. List the stations up the page consecutively from the beginning station on every fourth line. Enter the station of all transit points shown on the plans; such as, P.O.T. (point on tangent), P.I. (point of intersection), P.C. (point of curvature), P.T. (point of tangency), at the proper place in the field book according to the station numbers. The reference ties for each point should be entered opposite the corresponding station number on the right-hand page. The station number of each structure shown on the plans should be entered in the field book in order that centerline location of structure points may be established as the line is run.

It is advisable to plot the right-of-way lines showing right-of-way widths on each side of the centerline on the right hand page of the alignment book.

A transcript of each right-of-way deed should be available on the project to enable the checking of right-of-way as shown on the plans or the division of right-of-way should furnish a checked set of plans.

Computer data processing programs are available which can produce alignment stakeout notes. This information is produced from original design data, and can be printed out on convenient perforated paper that can be inserted in hard-bound notebooks. Data produced from these programs includes stationing, tangent bearings, curve deflections, central angle, radius, curve length, semi-tangent distance, deflection per foot, and external distance. This information can be sent directly to field crews for immediate use in staking out the project.

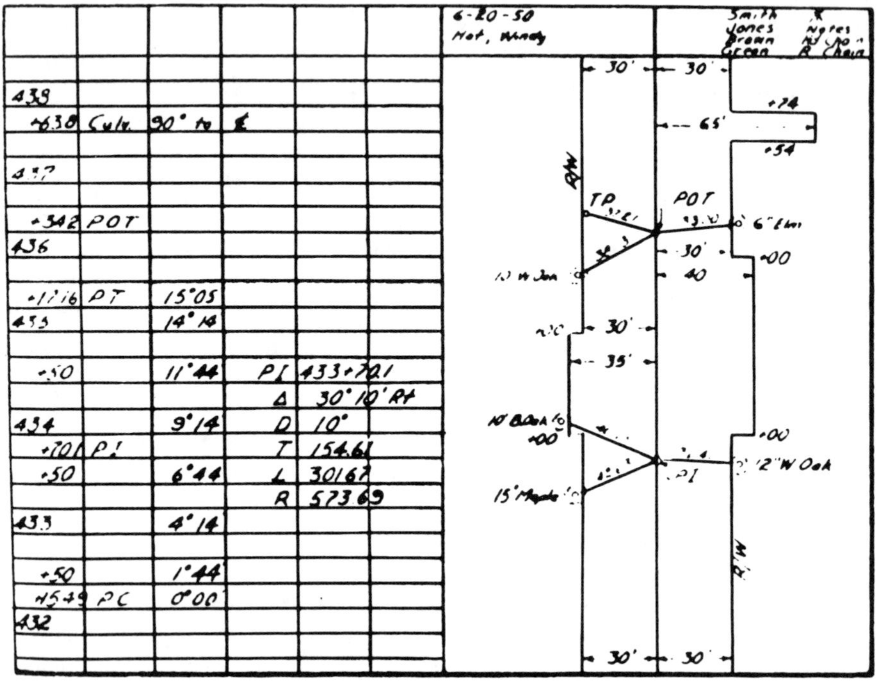

ALIGNMENT NOTES
Figure 150-1

C. Staking Centerline. Accepted survey methods are used in staking and establishing the centerline. All measuring is done on a horizontal plane. To compensate for sag, a pull is used for which the chain has been standardized, usually about 10 pounds.

Starting at the beginning of the project, as located from the nearest established transit point, each station should be set. Any pluses necessary for right-of-way breaks, curves, or for the centerline of each structure should also be set at this time. In establishing the centerline it is well to set a "red head" (spike and red cloth) at each station and each plus when a stake cannot be set.

Transit points shown on the plans should be located ahead of centerline staking. Small errors in chaining should be corrected at the first P.O.T. or at the first P.T. after the error is noted, and the resulting "equation" recorded if greater than one foot. If a very large error is found in chaining, the equation should be made as near the point of error as possible. An equation refers to a point on the survey that has two station numbers; for example, 185 + 60.3 (back) 185 + 24.7 (ahead). This is a plus equation of 35.6 feet. All equations shown on the plans must be observed.

When transit points are lost, or are not referenced, they should be set from known points that can be located. This can be done by producing tangents from established points or by the intersection of known tangents. All transit points set during the staking of centerline should be referenced to at least 3 objects. It is also well to set tacked hubs at right angles to the transit points along the right-of-way line on each side of the centerline. All ties for these points should be recorded for future use in the alignment book.

D. Simple Curve. Information on the computation of curve data can be found in a number of engineering handbooks; however, a brief explanation and example will be outlined here for the beginner. See Figure 150-2.

The curve data shown has been computed from the arc definition of degree of curve in which the radius varies inversely with the degree of curve. The radius of the 1-degree curve is set at 5,729.58 feet, therefore the radius of a 2-degree curve is 1/2 of 5,729.58 feet, the radius of a 3-degree curve is 1/3 of 5,729.58 feet, etc. For a given central angle the values of the various other functions of the curve also vary inversely with the degree of curve.

The angle of intersection at the P.I., generally shown on the plans as Δ, should be measured in the field as a check. The tangent length is computed by obtaining from tables the tangent length for a 1-degree curve, and dividing this length by the degree of curve. The length of the curve is computed by multiplying the angle of intersection (expressed in degrees and decimals) by 100, and dividing the result by the degree of curve (expressed in degrees and decimals).

Example, Given: P.I. — STA 33 + 70.1
Δ — $30° - 10'$ Rt.
D — $10° - 00'$

From table-tangent of 1° curve for $30° - 10'$ = 1,544.17 ft.
Tangent − T for $10° - 00'$ Curve = 1,544.17/10 = 154.42 ft.
Sta. of P.C. = (33 + 70.1) − (1 + 54.42) = 32 + 15.68
Length of curve − L = $100\Delta/D = 100 \times 30.167/10 = 301.67$ ft.
Sta. of P.T. = (32 + 15.68) + (3 + 01.67) = 35 + 17.35
Deflection angle = D/2 for 100′, D/4 for 50′, D/8 for 25′
Deflection per foot in minutes = $60D/200 = 0.3D = 0.3 \times 10 = 3'$

Ordinarily curves are staked for each station, half station, and quarter station. Normally the distance from the P.C. or P.T. to the first station on the curve is an odd distance and requires computation of the deflection for these distances. Chord corrections may be disregarded if 100 ft. chords are used for up to 6° curves, 50 ft.

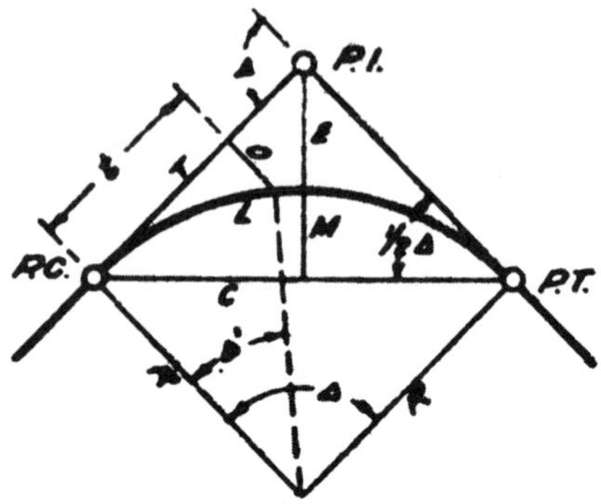

A 1° Curve is a curve of a radius such that 1° of central angle subtends an arc of 100′ in length

D	=	**Degree of Curve**
R	=	**Radius**
L	=	**Length**
T	=	**Tangent**
Δ	=	**Central or Deflection Angle**
E	=	**External**
M	=	**Mid Ordinate**
C	=	**Long Chord**
O	=	**Tangent Offset**
t	=	**Subtangent**
P.I.	=	**Point of Intersection**
P.C.	=	**Point of Curvature**
P.T.	=	**Point of Tangent**
E	=	$R \text{ exsec } \frac{1}{2}\Delta$
E	=	$T \text{ tang } \frac{1}{4}\Delta$
M	=	$R \text{ versine } \frac{1}{2}\Delta$
C	=	$2R \sin \frac{1}{2}\Delta$
O	=	$R - \sqrt{R^2 - t^2}$
t	=	$R \sin \Delta'$
R	=	$\dfrac{50}{\sin \frac{1}{2} D}$
L	=	$\dfrac{100\Delta}{D}$
T	=	$R \text{ tang } \frac{1}{2}\Delta$
Δ	=	$\dfrac{DL}{100}$
D	=	$\dfrac{100\Delta}{L}$

$$O = R - (R \cos \Delta') = R \text{ versine } \Delta' = R(1 - \cos \Delta')$$

CIRCULAR CURVE FORMULA

Figure 150-2

chords for up to 12° curves, 25 ft. chords for up to 24° curves, and 10 ft. chords for curves of over 24°. Curve notes should be recorded in the field book similar to sample page shown in Figure 150-1.

Computation of deflections:

Station	Dist.	Def.	
32+15.68 P.C.		0°00′	
32+50	34.33	1°43′	= 34.32′ × 3′
		+ 2°30′	= 50′ × 3′
33+00	34.32	4°13′	
		+ 2°30′	
33+50	50.00	6°43′	
		+ 2°30′	
34+00	50.00	9°13′	
		+ 2°30′	
34+50	50.00	11°43′	
		+ 2°30′	
35.00	50.00	14°13′	
		+ 0°52′	= 17.35′ × 3′
35+17.35 P.T.	17.35	15°05′	

Note: Last deflection is 1/2Δ

E. Staking Simple Curve. After setting the transit over the P.I., a backsight is taken on a P.O.T. on the back tangent and the tangent distance chained back from the P.I. to set the P.C. A foresight is taken on a P.O.T. on the forward tangent and the tangent distance chained to set the P.T. To establish the curve, with the transit set up over the P.C., sight the P.I. with the verniers of the transit set at zero. Turn off the deflection angle for the first point on the curve and measure the correct distance to this point. This point may be an even station or a station plus 50. After the first even station is reached the rear chainman holds on this station. The remainder of the curve is run in 50-foot increments with all chaining being from the full station points. The last deflection angle and the last measurement should both fall on the P.T. if no error has been made.

The curve may be backed in by using these same deflections. Set the transit on the P.T. with the total deflection set back on the vernier. Take a foresight on the P.I. and run the curve the same as before setting the recorded deflection for each corresponding station, finally tying in at the P.C. with the vernier reading 0 degree.

If the entire curve cannot be seen from the P.C. or P.T., it will be necessary to make an intermediate set-up on the curve. The first set-up will, in most cases, be at the P.C. and the curve will be run forward as far as practicable. The transit is then set up at the last point set and is backsighted on the P.C. with telescope inverted and the plates reading 0 degree (which is the angular value opposite the P.C. in the notes). After plunging the telescope, the angle opposite each succeeding station in the notes is used to set that station.

If still another set-up is necessary, then the transit is backsighted on the last transit point with the telescope inverted and the plates reading the angular value of the point sighted on, as recorded in the notes. The telescope is then plunged and the angle opposite each succeeding station in the notes is used to locate the stations.

F. Spiral Transition Curves. The purpose of introducing spiral transitions at each end of circular curves is to provide a gradual transition from the tangent to the full superelevated curve. The spiral transition has the property of ever increasing curvature, starting with zero degrees and ending with the same degree as that of the horizontal curve. Information as to spiral and superelevation transitions will be found on the standard drawings of the contract plans. Generally, only those curves of over 2 degrees in curvature and on major and interstate system routes will be spiraled. A complete treatise of spiral curves may be found in "Transition Curves for Highways" by Joseph Barnett of the B.P.R.

G. Staking Right-of-Way. After the centerline has been established, the right-of-way line should be staked on both sides of the roadway. Right-of-way stakes should be set at 50-foot intervals on curves. The staking should conform to the right-of-way as shown on the plans or as modified by executed deeds. Right-of-way lines for drainage easements and borrow pits should likewise be staked. Right-of-way stakes should be set at right angles to the centerline and should have the station number and distance from centerline plainly marked on them. Breaks in the right-of-way should be established by turning off with the transit the designated angle from centerline. It should not be necessary to actually turn 90-degree angles with the transit for other points unless the project requires wide and extensive right-of-way.

150.04 — Levels, Bench Levels, Cross Sections, and Slope Stakes

A. Office Work. Considerable office work should be done before field work starts. The grade line should be checked for errors, and grade elevations computed for each station and necessary plus. Widening, rate of superelevation, and transitions should be checked and listed in the field book for all curves.

B. Level Books. In preparing all level books, leave plenty of space between stations to avoid crowding notes. Except for identification of the field book and a list of benchmarks as taken from the plans, there is little to enter in the cross section book before field work starts. In the slope stake book, however, all of the profile grades and shoulder grades should be entered for each station and each plus. A list of benchmarks should likewise be carried in this book. On paving projects, a level book should be prepared in advance of setting paving stakes with the edge of pavement elevation shown for even and plus 50 stations.

C. Benchmark Levels. Prior to running any levels or establishing elevations for construction use, a set of benchmark levels should be run to check the elevations of benchmarks shown on the plans and to establish additional benchmarks. When checking bench levels, the length of foresights and backsights should be kept approximately equal and each benchmark should be used for a turning point. It is not advisable to accumulate small variations found in elevations of original benchmarks. In these instances, levels should be corrected before proceeding. See Figure 150-3 for an example of bench level field notes.

						6-21-50 Clear, Warm	Smith - ⊼ Brown - Rod
B.M.	BS(+)	HI	FS(-)	Elev.			
BM #1	4.72	951.44		946.72		USGS BM Brass knob in concrete base	
						125' Rt Sta 1+35	
T.P.	3.75	944.58	10.61	940.83			
T.P.	2.18	938.03	8.73	935.85			
B.M. #2	5.26	935.92	7.37	930.66		Spike in telephone pole, 45' Rt Sta 11+75	
T.P.	4.61	932.56	7.97	927.95			
T.P.	0.21	921.64	11.13	921.43			
B.M. #3	9.20	920.49	10.35	911.29		Spike in root of 14" Oak 65' Lt Sta 20+45	
T.P.	11.67	929.87	2.29	918.20			
T.P.	12.01	940.53	1.35	928.52			
BM #3	\		0.39	940.14		Hub with guard 100' Lt Sta 30+00	
Σ(+) =	53.61	Σ(-) =	60.19	946.72			
			-53.61	-940.14			
		Diff. =	6.58 =	6.58	✓		

BENCH LEVEL NOTES
Figure 150-3

At the completion of a project, additional benchmarks should be established on bridges, concrete box culverts, or other permanent objects. The new elevations and descriptions should be recorded on the final plans. Final benchmarks should not be more than 1,000 feet apart.

D. Preconstruction Cross Sections. Roadway common excavation is usually paid for on the basis of plan quantities; thus, preconstruction cross sections are not generally required. However, it is suggested that the original plan cross section be checked for accuracy by making a 3-point check while setting slope stakes, i.e., centerline, and right and left limits of excavation. If these points are found to be reasonably accurate, the original section may be used if a recomputation is necessary of the excavation quantities in any balance. If an unsatisfactory average variation is found in the original cross section, new shots should be taken and used in lieu of the plan section for computation of excavation quantities. The level rod should read to the nearest one-tenth of a foot and the distance from the centerline to the nearest one-half foot. The notekeeper, or designated person, should determine where any section is to be taken, and after some training, the rod person may select the individual shots along the section. Sections are to be taken as near right angles to the centerline as practicable. When a revision in excavation quantities is anticipated in a balance, or part of a balance, due to errors found in ground elevations, or change of grade or typical section, the Contractor should be so advised in writing by the Project Engineer during construction progress. See Figure 150-4 for an example of cross section notes.

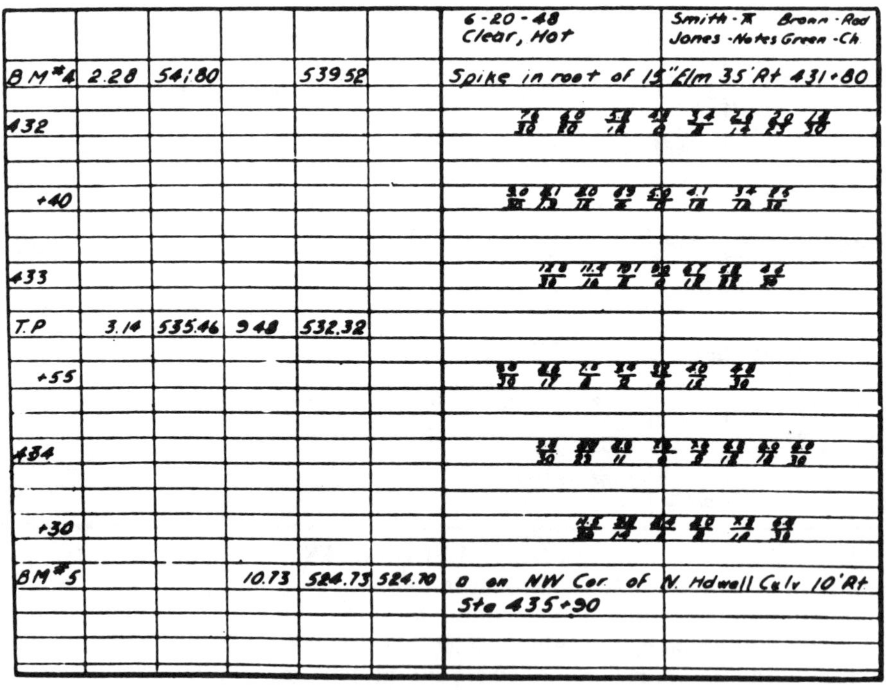

CROSS-SECTION NOTES
Figure 150-4

When rock is encountered, the rock area should be cross-sectioned as soon as the overburden has been removed. In case of extreme variation of the rock surface, such as pinnacle formation where a line of demarcation between rock and earth cannot be establaished, the District Construction Engineer, Project Engineer, and the Contractor's representative should agree as soon as practicable on a percentage factor of rock classification for this area. If sufficient time is available, the Division of Construction should be advised so that a general inspector may be present when determining the percentage factor. This method of classification is not considered good practice and should be used only when it is impossible to establish lines of demarcation. Any decision should be confirmed in writing.

Rock sections should be taken at the stations where original cross sections were taken, plus any additional sections that will be needed to arrive at the correct volume of rock excavation. After the rock has been excavated, and before any backfill is placed, level shots or cross sections must be taken to show the true lower limits of rock excavation. It is not necessary for the rock broken below the limits required for undergrading to be completely removed from the lower portion of the cuts prior to taking the level shots or cross section; however, the broken rock must be moved endways or sideways sufficiently to expose an area of at least one-half the width of the cut so that a true section can be taken. After sectioning this half, the remaining half must be exposed for sectioning in a similar manner.

While the above applies specifically to roadway excavation, the principles apply for channel changes and channel cleanouts where quantities have been computed. Where plan quantities are merely estimated, centerlines or base lines must be established and completion preconstruction cross sections taken. Since borrow excavation is not paid for on the basis of plan quantities, borrow areas must be staked and original cross sections taken before the Contractor begins work therein.

E. Final Cross Sections. As soon as practicable after final completion of any part of the work, final cross sections should be taken where they are required for the purpose of computation or recomputation of final pay quantities. These sections must be taken at exactly the same stations where preconstruction sections were taken. Each final cross section should have at least 2 "shots" at each end on the ground beyond construction limits so as to provide a common line between original and final sections. Explanatory notes, made when taking final sections, are often a great aid when plotting the sections later. Low-lying excavation areas which might be materially altered by high water or overflow (channel changes, borrow pits, etc.) should be cross sectioned immediately after excavation is completed.

F. Slope Stakes. Slope stakes are an important part of the staking job as they control the lines and grades to which the grading contractor will work. The gradeline, typical section, and roadway cross sections should be studied in order to be entirely familiar with the section involved before staking starts. A grade to which the road is to be constructed must be computed for each station, and each half station on sharp vertical or horizontal curves and in rock cuts. The grade line is a system of tangent lines with parabolic curves at the intersections. The tangent grade is shown as a percentage such as 4.50% which means 4.5-ft. rise or fall in each station. Going with the stationing, the ascending grades are marked plus, and the descending grade minus. The vertical curves are shown in lengths, as 200 ft., 300 ft.

Prior to computing vertical curve corrections, the tangent grades are checked by starting at the first intersection of tangent grades near the beginning of the project. Computation of grade elevations on a summit vertical curve with tangents of equal lengths are shown in the following procedure. A vertical sag curve is computed in a similar manner except that the offset is added to the tangent grade instead of being subtracted.

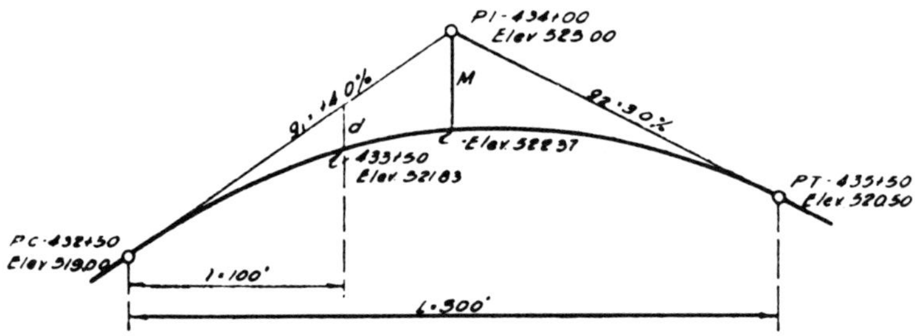

VERTICAL CURVES
Figure 150-5

$$\begin{aligned}
a &= \text{algebraic difference of tangent grades} = g_1 - g^2 \\
&= +\ 4.00 - (-3.00) = 7.00 \\
M &= aL/800 = 7 \times 300/800 = 2.625' \\
d &= 4M(1/L)^2 = 4 \times 2.625 \times (100/300)^2 = 1.17'
\end{aligned}$$

The tangent elevation at Sta. $433+50 = 519.00+4.00 = 523.00$
Curve elevation at Sta. $433+50 = 523.00-1.17 = 521.83$

The centerline, which is usually the profile grade, is computed on tangents where both shoulders are of equal elevation. On curves, the grades of the shoulders are not equal due to the superelevation. The amount of superelevation for the various curves is shown on a standard sheet in the plans. Elevations of each shoulder must be carried in the slope stake book along with any special ditches or variable slopes that may be designated on the roadway cross sections.

See Figure 150-5a for a sample of slope stake notes and Figure 150-6 for the location and marking of the stakes. In setting slope stakes the rod is read to the nearest 0.1 foot and the distances are measured with a metallic tape. All work should be checked for errors by comparing the slope stake distances as computed in the field against the cross sections as plotted in the plans or as shown on computer "Section Data".

Slope stakes should be set on all volume grading sections. They should be set at each station and at any other plus where they may be necessary due to curvature or

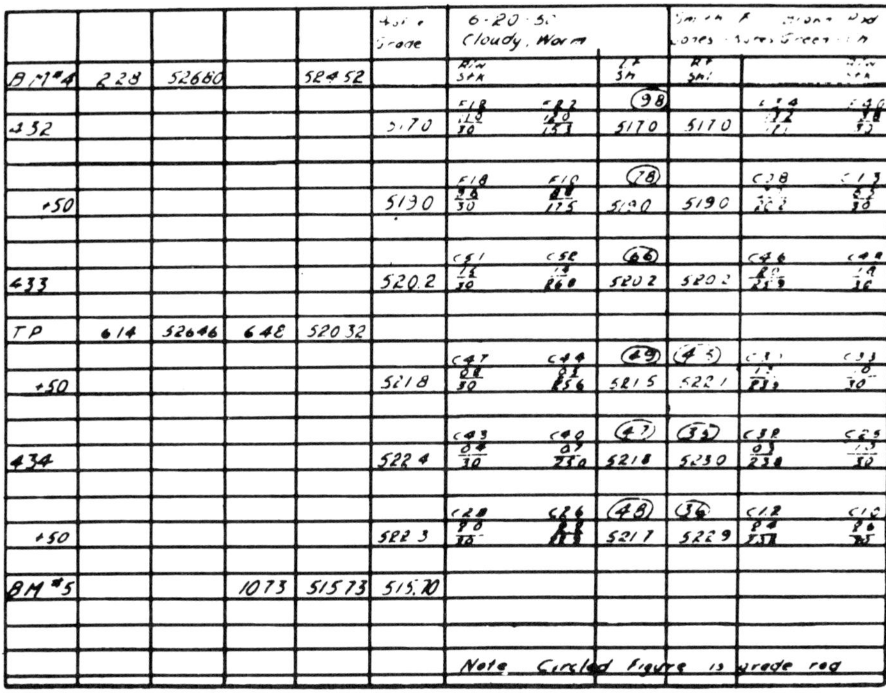

					[illegible] Grade	6-20-5[illegible] Cloudy, Warm			Smith [illegible] Jones [illegible] Green [illegible]		
						Rt Stk		Lt Sh	Rt Sh		Lt Stk
BM #4	2.28	526.80		524.52							
432					517.0	F18 / 10 / 30	F22 / 120 / 153	(98 circled) 517.0	517.0	C[illegible] / [illegible]	C[illegible] / [illegible]
+50					519.0	F18 / 98 / 30	F10 / 80 / 175	(78 circled) 519.0	519.0	C38 / [illegible]	C13 / [illegible] / 30
433					520.2	C51 / [illegible] / 30	C52 / [illegible] / 260	(66 circled) 520.2	520.2	C46 / [illegible]	C4[illegible] / [illegible] / 30
TP	6.14	526.46	6.48	520.32							
+50					521.8	C47 / 08 / 30	C44 / 03 / 256	(49 circled) 521.5	(45 circled) 522.1	C[illegible] / [illegible]	C3[illegible] / [illegible] / 30
434					522.4	C43 / 04 / 30	C40 / 03 / 250	(47 circled) 521.8	(35 circled) 523.0	C32 / 03 / 23[illegible]	C25 / [illegible] / 30
+50					522.3	C28 / 90 / 30	C26 / [illegible]	(48 circled) 521.7	(36 circled) 522.9	C12 / 84 / [illegible]	C10 / 86 / [illegible]
BM #5			10.73	515.73	515.70						
						Note Circled figure is grade rod					

SLOPE-STAKE NOTES
Figure 150-5a

rough topography. On sections of work that have varying slopes, it is well to indicate the slope on each stake in addition to other markings. Slope stakes should not be set on sections involving linear grading.

For channel changes and special ditches where there is a grade and typical section, slope stakes should also be set. Slope stakes may be necessary in borrow pits that are to be excavated to certain grades. Slope stakes are not always necessary for road approaches; however, where special grades are designed, they should be set.

Where rock is encountered, new slope stakes should be set on the top of rock for the rock slope as shown on the typical section or for the rock slope as determined by redesign. It is important that these stakes be set, as they will be used in the determination of overbreak quantities. Care should be taken to assure that the same proposed grade as was used for the original staking, except in the case of redesign, is used when setting subsequent lines of slope stakes.

A written order should be issued to the Contractor advising him of the method used in the setting of slope stakes and an interpretation of stake marking.

G. Paving Stakes. Stakes for pavement should be set with extreme care. The centerline must be reproduced to provide alignment for the pavement. Ordinarily a parallel offset line is run and the form stakes set at one side of the proposed pavement with alignment indicated on the stake with transit tacks. The offset distance should conform to the Contractor's request, if practicable, and is generally 2 or 3 feet from the edge of the pavement. There are several suitable methods of setting grades on paving stakes, however, driving the stakes flush and marking a cut or fill measurement from the top of the stake to the top of the paving form is generally preferable. If sufficient time and staking personnel are available, the driving of the top of the stake to form grade probably results in a more accurate form line. If it is necessary or desirable to give a cut or fill on the stake, it should be driven

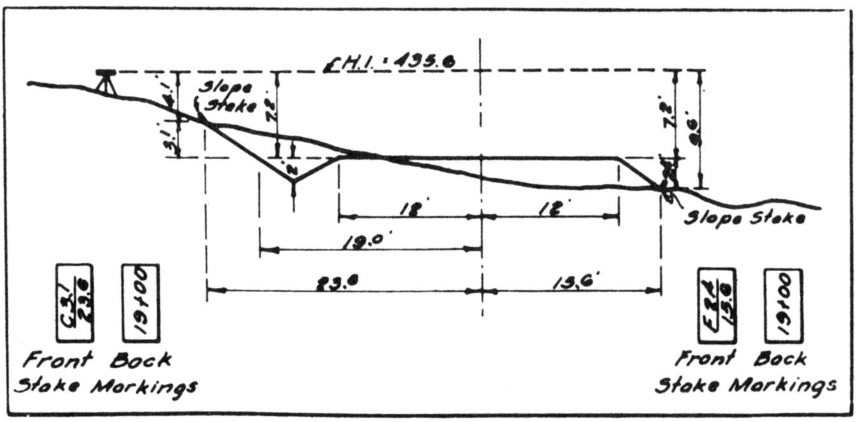

SLOPE-STAKING
Figure 150-6

so the grade may be marked to even tenths of a foot. Some engineers prefer to give both grade and tack line on the same stake. Paving stakes should be set at all stations and at each +50.

In some cases, due to abrupt alignment, steep grades, and spiral transition curves, it may be necessary to set them at closer intervals.

H. Finishing Stakes. On projects to be surfaced with a high type of surfacing not requiring the setting of forms, it is necessary that the grade of the roadbed be finished very close to the established grade and to the typical section width. For this reason, the setting of finishing stakes (blue tops) is required not only for the earth subgrade but for finished grade of the aggregate base. Finishing stakes are also needed on the grading work of major system projects where stage construction is being used with the paving to be done on a later project. On concrete paving jobs, most Contractors prefer to do their fine grade finishing after the paving stakes are set and, therefore, do not desire finishing stakes. If the Contractor requests finish stakes, however, they should be set. It is essential to set finishing stakes through rock cuts and to check the rock excavation before it is backfilled.

				Grades Lt Edge Pavt	Rt Edge Pavt	6-30-50 Clear, Warm		Smith ⊼ Green-Stks Brown-Rod	
BM #4	365	52817		52452			Lt Gr	Rt Gr	
432				51750	51750	10.27	10.67	10.67	10.47
						C 0.40			C 0.20
+50				519.50	51950	8.67	8.67	8.67	8.77
						Grade			F 0.10
433				520.71	520.71	7.36	7.46	7.46	7.26
						C 0.10			C 0.20
T.P.	7.64	530.35	5.28	522.89					
+50				522.96	522.63	6.57	7.57	7.90	7.00
						C 1.00			C 0.80
434				522.35	523.55	8.08	8.18	6.98	6.78
						C 0.10			C 0.20
+50				522.21	523.41	8.02	8.32	7.12	7.02
						C 0.30			C 0.10
T.P.	4.06	522.17	12.42	518.11					
BM #5			6.48	515.69	515.70				
						Note: Gr = Grade Rod			

PAVING STAKE NOTES
Figure 150-7

Finish stakes should be set at 50-foot intervals on the shoulder line and be driven so the top of the stake is at-grade elevation. It is common practice to blue the top of the stake to indicate that it is set to grade. Stakes or pins should also be set at centerline and on the quarter point where a close tolerance is required. It should not be necessary to set finishing stakes in linear grading sections for volume grading on supplementary construction.

I. Automated Slope Stake and Grade Books. Computer data processing programs are available which can produce slope stake and grade book notes. This information is produced from original design data, and can be printed out on convenient perforated paper that can be inserted in hardbound notebooks. The grade book program can be made to produce grades for subgrade, base material, surface, or any combination of lifts of material placed on the roadway. The slope stake program can be made to provide percent of grade, finished profile grade, vertical curve correction, crown, profile grade, subgrade shoulder, theoretical slope stake, cut or fill, and ditch notes. This information can be sent directly to field crews for immediate use in staking out the project.

J. Plans Prepared by Computer Methods. The earthwork quantities for some projects may be developed by electronic computer methods. In lieu of cross sections, the plans usually include forms showing such information as section data and yardage output data. Details will vary according to the computer and program used. Roadway cross sections will normally not be plotted except for areas requiring special consideration; however, culvert cross sections may be provided.

The computer form provides the same information normally shown in graphical form on the cross-section sheets of projects computed by conventional methods. Elevations for various control points of the section are usually tabulated with the offset distances from the surveyed centerline. The general slope of the existing ground can be determined from the elevation and offset of the slope stake points.

The same section data form may be used for both two-lane highways and divided four-lane roads. Details of the section must be determined from the typical section sheet included in the contract plans.

The outside shoulder points are frequently tabulated by elevation and offset. Inside shoulder points on four-lane divided highways can be established from the typical section. The ditch point offset is usually located at the toe of the slope. If a ditch berm is required, the berm elevation is listed.

The typical rock bench point identifies the elevation of the top of rock and the offset distances from centerline to the intersection of the rock backslope and the top of rock. The rock slope is shown on the typical section sheet.

The Quantity Output Data form provides the earthwork quantity information that is shown on the cross sections of projects computed by conventional methods. This may include end areas, volumes between sections, and accumulated volumes. The backslope may be shown as a number such as 7 = 1 – 1/2:1, 5 = 1:1, 3 = 3:1, and 2 = 4:1.

The slope stake points noted on the output data form are not to be used for actual setting of slope stakes but may be used as a check for determining if actual ground elevations vary from design elevations. If there is appreciable variation, a new cross section should be taken in the same manner as for plans designed in a conventional manner.

150.05 — Staking Structures

A. General. This section is devoted to the staking of bridges and culverts; however, the staking details of miscellaneous structures will be similar and should be worked out on the project. The staking of structures is an important part of the project work and a detailed study of the plans should be made so as to become thoroughly familiar with each structure before the staking starts. All bridges have separate plans where the dimensions, elevations, etc., necessary in staking the bridge are shown. An error in staking a structure will not occur if the staking procedure is thoroughly checked and some form of triangulation devised.

B. Field Notebooks. The field notebooks for structures should be prepared in advance and all necessary data placed in the book so it will be available when staking starts. Sample notebook pages are shown in Figures 150-8 and 150-9.

Culvert books should be prepared by using a separate page for each structure. This sheet should show the station of the structure, the type, size, and length. Staking diagrams, showing all dimensions and staked distances, should be made and a sketch showing the elevations of the flowlines included. This information is taken from the culvert plans which show the station, size, type, length, and the skew angle. Space for sufficient cross sections and elevations should be left to ensure properly computed structure excavation when same is required. The cross sections of the plans indicate the length of the culvert on each side of centerline and flowline elevations. If the culvert is a pipe, the length is to the ends of the pipe. The length of a concrete box culvert is to the inside face of the headwalls. The angle of the culvert is shown as 90 degrees if at right angles to the centerline, or a certain degree skew from right angles if other than 90 degrees. The skew is written RA or LA, meaning right ahead or left ahead.

Bridge field notebooks are prepared in a manner similar to that used for culverts. A separate page is used for each bent of the structure, with the staking diagram showing the required dimensions and staked distances. A sketch showing an elevation of the bent together with the critical elevations should be included. The left-hand page is to be used for recording the required leveling data.

C. Pipe Culvert Staking. With the transit set over a point (tacked hub or nail) on centerline of survey at the station of the centerline of culvert, take a sight on a survey control point (P.I., P.C., P.O.T., etc.) and turn the angle for the centerline of culvert. Generally 2 hubs to establish the centerline are all that are required for a pipe culvert. Set one at each end, offset a definite distance (5 feet minimum) from the ends of the proposed pipe. If there is a break in the flowline, set 2 hubs, right and left of culvert centerline at the break to mark its location. From the nearest benchmark determine the elevation of the top of each hub and compute the cut or fill from it to the respective flowline point. Back of each hub set a guard stake on which has been marked the hub's identity, offset distance, and the cut or fill to flowline elevation.

The culvert as designed should be checked after staking to see that it fits existing conditions. In most cases it will be found that the culvert as planned will function satisfactorily. It is not intended to encourage habitual revision of culvert plans and when occasionally a revision is a necessity, it should be made as early as possible. Take sufficient cross sections for the recomputation of excavation quantities if justified by a change in location or flowline elevation. Usually a single section along the culvert centerline is sufficient, but if additional sections are necessary, they should preferably be taken parallel to the culvert centerline.

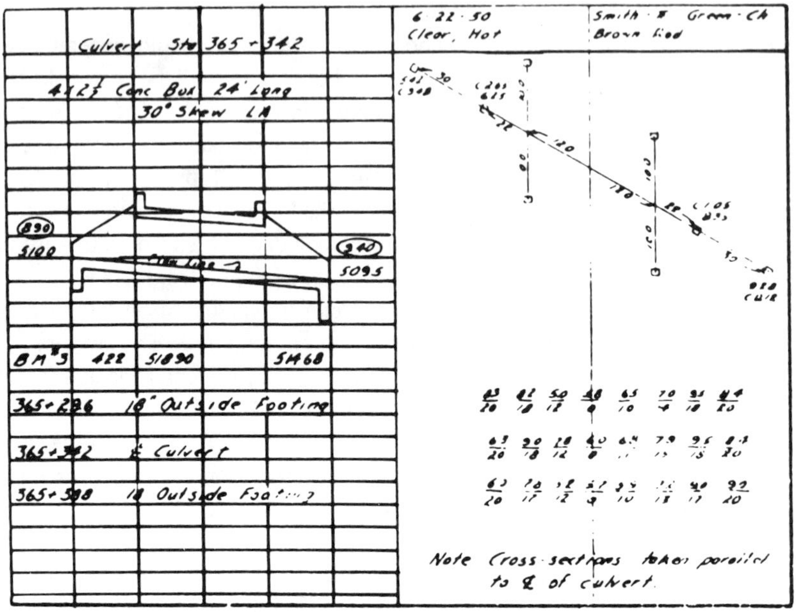

CULVERT STAKING NOTES
Figure 150-8

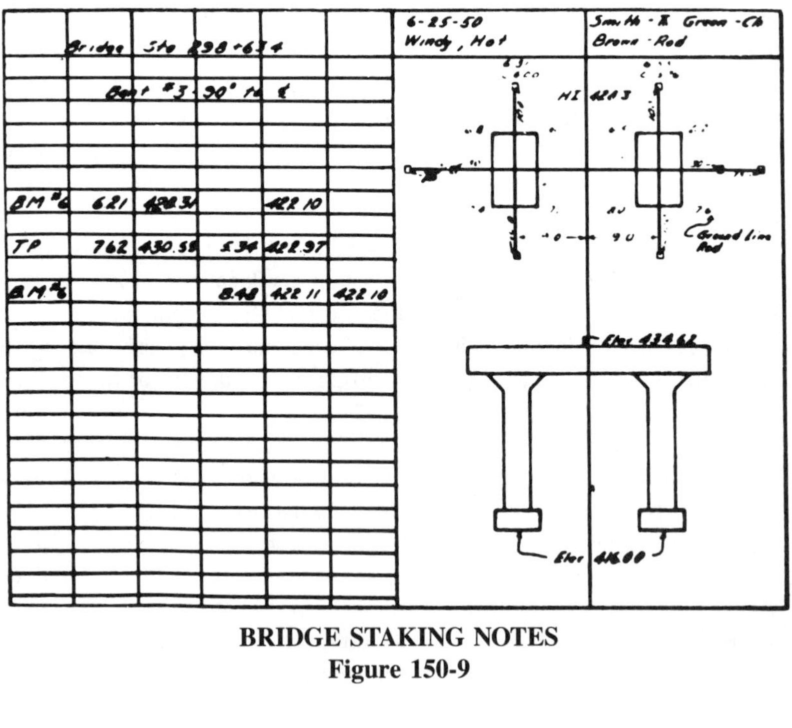

BRIDGE STAKING NOTES
Figure 150-9

D. Concrete Box Culvert Staking. The same method is used to stake concrete box culverts as for pipe, except, in addition, stakes must be set for the inside face of each headwall which are usually parallel with the centerline of the roadway. To set headwall stakes, a point is set along the centerline of the culvert at the inside of headwall distance; the transit set over this point and a sight taken along the centerline of the culvert; and the transit turned the same angle as turned at the centerline. The headwall line should be checked visually to make sure it is parallel to centerline of roadway. Two stakes marked "inside headwall" are set well back from the construction limits, one on each side of the culvert with a tack for line.

Box culverts as designed should be checked against existing conditions, as discussed under "Pipe Culverts," keeping in mind that any change should be avoided if possible. In case a revision of flowline elevation appears warranted, it should be ascertained that no commitment have been made concerning flowline elevations to various agencies or individuals before any changes are made on same. A culvert classed as a bridge should not be changed in any respect without prior approval of the Division of Construction and the Division of Bridges.

1. **General.** The station numbers shown on the plans for a bridge are the stations at the ends of the bridge, usually at the fill face of the end bents. Set the transit on a P.I., or P.O.T., near one end of the bridge, and take a foresight on a P.I. or P.O.T., beyond the bridge. Set at least 2 substantial tacked hubs well past each end of the bridge, preferably on high ground that will not be disturbed. This gives 4 points on the centerline of the bridge. If one is lost or destroyed, the line may still be reproduced. From the station of the bridge, the location of the first bend or abutment can be determined. The work line for the end bents is generally the fill face. Set a tacked hub on centerline at this point. Measure from this point the distance to the centerline of the next pier or bent and set another tacked hub. Continue this until you have tacked hubs at each pier, bent, or abutment. Set the transit over each of these points and stake the working line by turning the angle shown on the plans. At least 2 tacked hubs on each side of the bridge, clear of all construction operations, should be set for each pier, bent, or abutment. With the transit still set over each of the points, measure the distance out of centerline of footing and set a point. Set the transit over this point and turn the centerline of each footing. Set tacked hubs on each side of the footing by turning the angle indicated on the plans, which will usually be 90 degrees. Repeat this operation until each pier, bent and abutment is completely staked.

2. **Triangulation.** Bridges over wide expanses of water will require the use of triangulation to fix the location of river piers and to determine the exact stationing of the substructure units located on the landfall ahead. This triangulation should be accomplished in the manner set forth in any survey handbook.

3. **Benchmarks.** A good benchmark that will not be disturbed should be established at each bridge site. The elevation of this benchmark should be thoroughly checked and it should be used to set all future control elevations at the bridge. After the staking is complete, lay out each footing approximately and obtain

sufficient elevations of the ground so the structure excavation may be computed. Grades should be set to show the elevation of the bottom of the footing as noted on the plans. A guard stake should be set by each hub on which has been written complete accurate information regarding the hub; what point or line it marks, its offset distance, cut to grade, etc.

4. **Checking.** A thorough check of all measurements, angles, and elevations should be made to make sure no errors exist. The checking should, if possible, be done by persons different than those who did the original staking. If possible, they should also use different methods. One method of checking and staking of the centerline of piers would be to compute the station and plus to the nearest one-hundredth foot for each pier and then locate each unit by stationing. A further check could be made by measuring the overall distance between end bents to assure that it equals the sum of the individual bent distances. A check on the skew angles turned can be made by measuring on a line parallel to centerline through the offset stakes for each pier. Always be positive the staking is right—never take anything for granted when staking a bridge. If the least doubt arises, consult the plans. Check the dimensions on the plans for errors. Normally, they are correct but do not take it for granted that errors are never made on the plans. If errors are discovered, the matter should be referred to the designing agency for correction.

150.06 — Measuring, Clearing and Grubbing

The specifications set out the method of measuring, clearing and grubbing units. Generally, rectangular units are specified, and are staked along the centerline of roadway. When trees or brush that qualify under the specifications as clearing are encountered, the number of units must be determined. A note-keeper and two rod carriers can do this. Starting at the centerline, the width of each unit is laid out and only those with clearing in them are counted. Consideration must be made for those areas that are marked on the plans to be undisturbed. At this time, it can be determined which of the clearing units are to be grubbed, based on the proposed cut or fill in each unit. A complete record should be made in the field notebook of the count of units. The Contractor's representative should be in attendance when the check for the record is being made and his/her signature in the field book is requested to indicate the Contractor's agreement of the count. If the Contractor's representative does not care to be in attendance, this fact must be documented by a memorandum from the Contractor.

It is the policy of the Department to retain a limited number of trees or shrubs to blend the right-of-way into the adjacent terrain. The trees to be saved should be marked with a tag, red cloth, or other suitable identification to indicate to the Contractor's personnel that the tree is not to be cut. Care should be exercised in selecting the units to be retained so that only sound, vigorous trees or shrubs of desirable species be saved in locations that will not create future maintenance problems or safety hazards.

150.07 — Land and U.S. Survey Monuments

A. General. It is important that all land monuments of section corners and subdivision corners and all U.S. Government monuments within the limits of the right-of-way be preserved. These marks are for the use of all engineers and surveyors, and the policy of the Department is to cooperate in their preservation. An examination of the plans should be made for location of section lines or subdivision lines crossing or intersecting other such lines within the limits of the right-of-way. The plans may also indicate the land monuments. Office notations should be made in the field notebooks of indicated monuments and land line intersections so the locations may be investigated for existing monuments.

B. Land Monuments. All monuments that will, or may, be disturbed by construction operations should be carefully referenced in the presence of the County Engineer or a representative, if possible. The County Engineer should be notified in advance, and every effort made to ensure the Engineer's presence when the monument is referenced. There should be at least 3 substantial tie points sufficiently removed so there is no chance of their being destroyed during construction operations. When no monument is visible, investigation should be made to determine positively that it is not buried.

Prior to replacing monuments, the County Engineer should be advised of the date proposed for resetting. Again, an effort should be made to ensure the Engineer's presence. When on a graveled road, the monument should be set about 6 inches below finished grade. If on a paved road, a metal marker, such as a railroad spike, should be inserted flush with the surface of the pavement, and a cross chiseled or filed thereon to show the corner. Careful notes should be kept of the references, and the notes should be witnessed by two or more persons. Upon completion of the project or as soon as monuments are reset, a letter should be written to the County Engineer, with a copy to the District Office, setting forth exactly what was done. It should include a description of all corners referenced, description of all reference points, and names of all witnesses.

C. U.S.G.S. Monuments. U.S. Geological Survey monuments may serve as both a level monument and traverse monument, or for each one separately. In the event it is necessary to move one of these monuments, immediately notify the U.S. Department of the Interior, Geological Survey, located in the National Center, Reston, Virginia, 22092, phone 703-648-4000. In the notification, give a general description of the area in which the monument is located, together with the designation as stamped on the bronze plate. The U.S.G.S. will then either send a party to move it or send forms on which to enter the description of the new location and field notes for the revision of data. In case of level monuments which are moved by project forces, the difference in elevation only need be recorded with proper field notes to substantiate this difference. To preserve the accuracy as nearly as possible, the difference in elevation should be given to the thousandth of a foot. For a traverse monument, the data to be determined is the true bearing to the nearest 10 seconds, with horizontal distance to the nearest one-tenth foot from old to new location. Where the true bearing is not available, reference the monument with some accurate transit ties so the exact location can be reestablished. Send this information to U.S.G.S. so they can reset the monument with their own forces when a party is available.

D. U.S. Coast and Geodetic and National Geodetic Survey Monuments. When a NGS (formerly U.S.C. & G.S.) monument is found and will be disturbed by construction, immediately notify by letter the U.S. Department of Commerce, the Director, National Geodetic Survey N/CG162, Rockville, Maryland, 20852, phone 301-443-8316. A copy of this letter should be sent to the U.S. Department of the Interior, Geological Survey, and a copy to the District Office. This letter should give the reason for requiring the monument to be moved, a general description of the area in which it is located, and the designation data stamped on the bronze plate. If it is urgent that the monument be moved at once, request an early reply to the letter. Full instructions will be sent for resetting the monument. After the monument is reset, send the requested data to the National Geodetic Survey.

150.08 — Right-Of-Way Markers

Right-of-way markers should be set on all projects. If they are not an item in the contract, then they should be set by State forces. In the latter case, it is suggested that the markers be set when the project is being staked for construction, or as soon thereafter as personnel is available for the work.

Section 180 — Utility Relocation and Adjustment

180.01 — General

Frequently, when the installations of utility companies are affected by the construction or improvement of a highway, the utility companies are obligated to move, relocate, and protect such portions or sections of their installations as necessary to facilitate the construction or improvement of a particular project. In some instances, the Department may reimburse the utility company for the cost of such work; in others, it may be necessary for the utility to bear the cost of such work.

Where the cost of the necessary relocation and adjustment of such utility facility is an obligation of the Department, a utility agreement is generally negotiated between the Department and Utility Company for performing relocation.

In some instances, this work is accomplished by including the necessary items of work in the general contract on a unit price basis. In other instances, this work may be accomplished by a lump sum price or an actual cost basis between the Utility Company and the Department.

180.02 — Definitions

Utility shall mean and include all privately, publicly, or cooperatively owned facilities consisting of electrical energy, oil, gas, water including sewer, steam and other pipe lines, and communications facilities. Dependent upon the meaning intended in this manual, the term "utility" may also mean the utility company, inclusive of any wholly owned subsidiary.

The term **reimburse** shall mean that State or Federal funds may be used to reimburse the utility to the extent provided by law.

Cost of right-of-way shall mean the costs of land required for the relocation of the utility facility.

Preliminary engineering shall mean and include locating, making of survey, and the preparation of plans and estimates prior to execution of agreement.

Construction shall mean the actual building and all related work including utility relocation or adjustments.

Salvage value is the amount received for utility property removed, if sold, or if retained for re-use, the amount at which the material recovered is charged to the materials and supplies account.

State as referred to herein shall mean the Departments of Highways or Transportation.

Authorization shall mean authorization to the utility to proceed with any phase of a project, by the State.

Betterments shall mean and include any increase in the functional capacity of or service improvement in the facility being relocated or adjusted except to the extent necessitated by the highway construction.

Overhead costs shall mean those costs not chargeable directly to accounts pertaining to the relocation which are determined on the basis of a rate or percentum factor supported by overhead clearing accounts, or such other means as will provide an equitable allocation of actual and reasonable overhead costs to specific relocation jobs. Such costs may include expenses for general engineering and supervision, and general offices services, relocation engineering and supervision by other than the accounting utility, legal services, insurance, relief, pensions, and taxes.

180.03 — Preconstruction Conference

Following the award and execution of the highway improvement contract, representatives of the utility companies and other affected and interested parties should attend a preconstruction conference.

180.04 — Progress of Work

The notice to proceed with the work is given by the District or Project Engineers or Utility Division representative in charge of the construction. In the event that the work under a utility agreement is performed prior to the start of operation under a highway improvement contract, the Project Engineer or Utility Division representative will assign an Inspector to supervise the work under the utility agreement.

The Project Engineer or Utility Division representative through the Inspector keeps a record of the progress of the utility adjustments.

180.05 — Inspection of Work

An inspection should be made of the utility adjustments to make sure the facilities are located outside the limits of the planned highway improvement.

It should be determined that the utility forces and project personnel use the same reference datum when setting grade stakes, and confer with Project Engineer or Utility Division representative prior to establishing any underground, overhead, or lateral installations.

180.06 — Contract Records

A. General. Daily construction records should be maintained in order to ascertain that the proposed work is accomplished in accordance with the requirements of the agreement and to furnish the evidence necessary for recommending payment to the utility company.

B. Actual Cost or Force Account Agreements. For this type of agreement, the Inspector should keep a daily record of the number and classification of personnel employed, the material used, major items of equipment used, and any other information that may be of assistance in verification of billing charges. A record should be made of all materials removed from the job site that are returned to stock or scrapped. These daily records should be certified by the Inspector.

In the event the utility company has the work performed under an existing continuing contract on a unit price basis, daily records of personnel and equipment are not essential, but a daily record should be made of the work operations by stations and the number of units of work completed. However, if the utility has the work performed by an existing continuing contract on a force account basis, daily records should be made of personnel and equipment, in the same manner as prescribed for work performed by the utility on a force account basis.

C. Unit Price Basis (General Roadway Contract). When the utility relocation is performed by the General Contractor or agents, a daily record should also be made of the work operations by stations and the number of units of work completed. This should be included in the Inspector's diary for each day that work is performed.

D. Lump Sum Agreements. When a utility relocation is performed by a utility company under a lump sum agreement, daily records are not required of hours, material items or equipment time, but the Inspector should assure that the work is accomplished in accordance with the requirements of the agreement and should keep sufficient records of work performed to enable certification that the work has been accomplished in the manner prescribed in the agreement.

E. Minor Changes in Approved Work. Utility companies may be authorized by the Project Engineer to do all work involving minor changes in quantities or minor items not included in the approved estimate that may be necessary to accomplish the intent of the approved utility agreement. This action may be taken without necessity of formal approval, with the understanding that the Project Engineer's or Utility Division representative's record and final billing will provide adequate documentation of such minor changes.

Substantial changes in the scope of work covered by the utility agreement and/or substantial change in location must be submitted for approval by the State and on Federal-Aid work by the Federal Highway Administration either verbally or in writing and will, if necessary, be reviewed in the field. If such proposed changes are found to be satisfactory, verbal authorization for such changes will be given and confirmed by letter. Prior to the time of final billing, revised preliminary plans will be required in order that the State's files will reflect the true location of the relocated facility.

Minor or substantial changes in work may necessitate revision of betterment or extended service life percentages established at the preliminary stage and agreed upon as being applicable to the final billing.

F. Inspection of Recovered Materials. The purpose of inspecting recovered material is to prevent the junking or scrapping of all recovered material and to assure that proper allowance is made for any salvaged material.

On Federal-Aid work, the Federal Highway Administration should be notified by the Department by letter of the time and place where the recovered materials to be sold or scrapped will be available for inspection. This letter should show a breakdown of the materials involved. This will prevent any possibility that the utility company may be cited for the salvaged value of the materials when the audit is performed.

A statement is required on the final bill that credit has been allowed for all salvaged material and that items scrapped were available for inspection and proper notice given.

G. Procedure to be Followed to Obtain Permits for the Installation or Maintenance of Facilities Within Limits of a Highway System. No utility facility shall be initially installed over, under, or along the State highway system without the utility company first applying for and obtaining a permit authorizing such installation. No deviations from this policy shall be permitted.

The application for this permit shall be prepared as required and must provide sufficient information so that the location of the proposed work may be easily found. A plan must be attached to each copy of the application. In addition to describing the general location of the work, the plan must accurately show the distance of the proposed facility from the edge of the shoulder or other limiting factors defined by the minimum standards.

A joint field inspection of proposed utility installations within State highway right-of-way limits, subsequent to the completion of the utility company survey and application, but prior to the issuance of the permit may be required either by Department or Company representatives.

The utility company's work should be inspected as it is in progress, and also after the work has been completed. Daily records should be kept of the work for each day that work is performed and should be made a part of the permanent records.

DIVISION 200 — EARTHWORK

Section 201 — Preparation of Right-of-Way

201.01 — Description

The initial operation to be undertaken in the construction of a highway facility is the preparation of the right-of-way and any designated easements. All obstructions and objectionable materials are to be removed and disposed of in accordance with the governing specifications. Items to be removed may include trees, undergrowth, stumps, rubbish, buildings, foundations, abandoned utilities and drainage structures, and any other materials or objects not needed in the construction, or of an unsuitable nature.

Construction Practices

201.02 — Protection and Preservation of Property

The Contractor is responsible for the protection and the preservation of all property, both in and adjacent to the right-of-way. The plans and specifications will provide the detailed instructions for clearing the work site, but in general the following should be considered.

A. The boundaries of all areas in which the Contractor will be required to perform work should be clearly marked. It should be made clear to the Contractor what ingress and egress has been provided for access to the project and for those property owners adjacent thereto.

B. This is the proper time for the Project Engineer and the Contractor (or their representatives) to travel over the job and discuss the general work to be done and any special details. Such details could include trees to be saved, stakes and survey control points to be preserved, environmental and archaeological comments, and all known utilities that could be damaged during clearing operations. The general clearing procedure and disposal of materials could also be discussed.

C. A careful investigation should be made to determine the location of all underground facilities within the right-of-way. With adequate knowledge of their location and depth, they can be protected against damage during the Contractor's clearing and excavation operations. Prior to beginning any clearing operation this information should be in the hands of the Contractor in order that their location can be established on the ground. There are occasions when utilities are in place at the time the contract is awarded that are to be removed or relocated within the right-of-way after certain construction operations are complete. The Contractor must be kept informed as to the intended time of their removal or relocation so that the construction operations can be properly coordinated with the activities of the utility company involved.

D. All trees, shrubs, survey or historical markers, objects of historical or archaeological value, etc., that are to be preserved or remain in place must be clearly marked and the Contractor made aware of their location.

E. Private or public property adjacent to the right-of-way, and all natural growth and improvements thereon, must be continually protected from damage by the Contractor's equipment and any construction operation that is underway. When or where such damage is done by the Contractor's equipment and operation, it shall be the Contractor's sole responsibility to rebuild or make good such damage or injury at his/her expense. The manner in which a Contractor preserves and protects such property should be of his/her selection as long as it is reasonable and consistent with good construction practice.

201.03 — Clearing and Grubbing

Excavation Areas

A. Clearing and Grubbing. All trees, undergrowth, stumps, shrubs, etc., not designated to remain in place, shall be cleared and grubbed. The choice of equipment is usually left to the Contractor as long as the equipment used will perform the work in a satisfactory manner. The use of explosives in congested areas should be controlled. The methods used to clear timber may vary from simply sawing a tree down to complete removal of tree, stump, and attached roots by pushing it over and rooting the stump out of the ground with a bulldozer.

B. Preservation of Trees. Trees and shrubs which will not interfere with the use of a highway and its drainage system are often selected to remain in place for their scenic, historical, or other value. These must be left in place undamaged. Therefore, some precautionary measures must be taken in order to protect these trees from damage during the clearing and other construction operations. Clearly marking these trees will help preserve them. It may be required that low-hanging and unsound and unsightly limbs be removed from those trees that remain in place. This must be done in a manner that will not damage the remainder of the tree. Branches of trees extending over the roadbed should be trimmed to give a clear height of 20 feet above the roadbed surface. All trimming should be done by skilled workers and in accordance with good tree surgery practice.

C. Extent of Removal. Trees, stumps, and large roots should be removed from excavation areas for roadway, channel, etc., to a depth that will satisfy the specifications and prevent such objectionable material from becoming mixed with the material being incorporated in the embankment. These areas to be excavated with normally require scalping to remove small bushes, vegetation, rubbish, and other objectionable material. In addition, areas of heavy timber, undergrowth, and small trees may require root raking to remove the majority of the roots left in the ground after the aboveground portions are cleared.

Embankment Area

A. Clearing and Grubbing. All trees and undergrowth shall be cleared and grubbed unless noted on the plans or in the specifications to be done otherwise. The general requirements listed under "Excavation Areas" will usually apply to areas being prepared for embankment.

B. Exceptions. The plans and specifications may allow the trees to be cleared only in areas to contain embankments. That is, the trees and any existing stumps need not be grubbed out but can be cut off as close to natural ground as possible and left in place. This is normally allowed only in areas where the undisturbed stumps are sound and will be a minimum depth below the subgrade or side slopes of the embankment as set forth by the specifications.

C. Maintenance of Ground Surface. All holes, ruts, etc., remaining after clearing and grubbing, should be backfilled and compacted as directed by the Project Engineer to prevent ponding of water. The cleared areas may also require blading to improve drainage wherever possible.

Disposal of Materials

A. Merchantable Timber. Timber, where required by the plans and/or specifications to be disposed of as saw logs or as pulpwood, should be trimmed of all limbs and cut into appropriate lengths. It is then either stacked out of the construction area or disposed of as directed by the governing specification. In many cases the timber of merchantable quality is not distinguished from other timber and becomes the property of the Contractor. The Contractor should be encouraged to market as much wood as possible to avoid other disposal problems.

B. Burning of Timber. In areas where burning is not prohibited because of fire hazard or regulations or where smoke will not present a problem, the combustible materials such as trees, limbs, stumps, and brush are normally burned in the open within the construction limits. Burning within the right-of-way should be under the constant care of competent guards and accomplished in such a manner that anything designated to remain in the right-of-way, surrounding timber or grassland, and all other adjacent property will not be jeopardized. All burning must be done in accordance with the laws and regualtions of Federal, State and local government agencies having jurisdiction.

C. Burying Timber. To avoid the labor and equipment cost of burning or otherwise disposing of clearing debris, the Contractor may, if approved or authorized in the contract, bury it within the right-of-way, disposal areas established by the plans, or other locations approved by the Project Engineer. Unauthorized burying can be prevented by constant observation of the Contractor's operation and by close examination of all ravine bottoms, etc., to see that they are clean and ready to receive embankment.

D. Disposal of Material Outside the Right-of-Way. The Contractor may elect to dispose of all, or part, of the clearing debris outside the right-of-way. Unless disposed of at a licensed waste disposal site the Contractor must have written permission from the owner of the land on which the disposal is made. Proposed disposal areas should be examined and approved by the Project Engineer after a determination has been made that they will not present an unsightly appearance, block natural drainage, or create the possibility of damage to adjacent property. A disposal site may require an archaeological clearance and the Project Engineer should be satisfied as to this need before granting permission to the Contractor. The Project Engineer should ensure coordination with environmental commitments and permit regulations regarding wetlands, 404 permits, etc.

201.04 — Hazards

A. Blasting. Blasting operations should always be conducted under careful, competent supervision to prevent damage to adjacent property and injury to persons. The location of the area in which blasting is to be done will determine, to a great extent, the procedure to be used. Both the size of the area to be shot at one time and the strength of the charges set therein should be determined by this factor. Blasting charges, as a general rule, should be as light as possible, yet of sufficient strength to adequately break up the material being shot. In restricted or urban areas where the hazard to surrounding life and property is at a maximum, it may be necessary to load fewer, and closer spaced, holes with lighter charges to break the material while holding the possibility of damage to a minimum. Blasting areas must be properly signed to inform all persons of potential hazard in the immediate vicinity of explosive charges. All explosives must be stored in a secure manner in compliance with local law and pertinent safety regulations.

B. Tall Timber. Clearing and grubbing operations, particularly in dense, tall-timbered regions can be very dangerous. The engineering personnel on the project should be especially careful when working in the vicinity of clearing operations. The Contractor should make certain that the operations do not endanger the engineering personnel, result in potential hazards to the traveling public, or create damage to existing facilities in or adjacent to the right-of-way. Extremely tall timber demands extra care and the proper equipment to get it on the ground without damage to surrounding property.

C. Fire. The type of area being cleared and the adjacent development should determine the manner in which burning is accomplished. Serious fires outside the right-of-way can develop if the Contractor loses control of the burning operation. Improper cleanup of the area surrounding the burning pile, a lack of guards to control the spread of fire, or smouldering timber rekindled by high winds are a few of the many causes of fire spreading beyond the intended burning area. There are many locations where it is advisable to use chippers or to restrict the size of the piles or select more favorable conditions of weather and ground moisture before burning is started. Burning should never be allowed near or under utility lines. The piles should also be located within the cleared areas of right-of-way in a location that will prevent the spreading of fire to adjacent timber or other combustible materials. Adequate, basic fire-fighting equipment should always be available.

201.05 — Records

A daily record of events should be prepared by the Project Engineer or Inspector in charge of the work. Items should be included as follows:

A. The limits in which work was accomplished for that day.
B. The type and number of pieces of equipment used in the work.
C. The number and classifications of labor used.
D. Discussion of project prosecution with the Contractor's representative which are of an unusual nature. If a discussion of a problem leads to specific recommendations or instructions being given the Contractor, record it.
E. Weather for the purpose of determining a working day.
F. Damage to private property caused by the Contractor's equipment or the operation should be noted and action taken recorded.
G. The final disposition of salvable materials should be noted.

201.06 — Measurement for Payment

Project records must be kept in such a manner that all partial progress and final payments can be easily and clearly supported by recorded data. This data should include when the work was completed, measurements and calculations to support the quantity allowed, and should be initialed or signed as required by the persons who made the calculations and measurements. It is recommended that a continuous record be kept, immediately subsequent to their completion, of the items that are paid for as each, linear foot, square yard, etc., and this record must include measurements made to substantiate the quantity allowed.

Section 202 — Removal of Structures and Obstructions

202.01 — Salvable Materials

All designated salvable materials should be removed, without unnecessary damage, in sections or pieces which may be readily transported, and shall be stored by the Contractor at specified locations. Steel structures that are removed, and required to be salvaged, shall have matching components clearly marked prior to disassembly. A copy of the match-making plan should be furnished by the Contractor for later use.

Section 203 — Excavation and Embankment

203.01 — Description

The construction of a graded roadbed, upon which the base and wearing courses will be built, is generally referred to as earthwork. Excavation is that part of the earthwork that is dug up, hauled, and then used to form the embankment portion of the roadbed. Roadway excavation, which is that material obtained from within the right-of-way, exclusive of channel and structural excavation, may be composed of common earth, solid rock, loose rock, or any combination of these materials. When there is not enough suitable roadway excavation, together with suitable excess channel and structural excavation, to complete the embankment to the required line and grade (i.e., the "cut" and "fill" do not balance), then borrow excavation must be imported to make up the difference. Under certain conditions it may be necessary to obtain the excavation from under water (lake, swamp, etc.) and in this operation the embankment may be completed by using a hydraulic dredge to excavate and move the material into place.

203.02 — Protection of Property

This is a continuation of those practices initiated during the phase of right-of-way preparation. There will be less restriction to the movement of equipment during excavation, and subsequent embankment construction, because the right-of-way has been cleared and grubbed. However, there still remain in many cases utilities (both above and below ground), trees to be preserved, survey control points, etc., that must be protected during this operation. There is also private property adjacent to the right-of-way that must be protected from damage by the Contractor's equipment. Prior to beginning this operation, it is advisable to again emphasize protection of property to the Contractor.

203.03 — Excavation (General)

Excavation involves the loosening, digging, loading, hauling, and disposal of material obtained from roadway cuts, channel changes, ditches, structure foundations, and/or borrow pits. Disposal of the materials is accomplished by incorporating it into an embankment or wasting that amount that is surplus or of an unsuitable nature. This item of work includes the constructing, shaping, and finishing of all earthwork for the entire length of the roadway, and approaches to same, in conformity to the required lines, grades, typical sections, and pertinent specifications.

If subsurface plans or information are available, the Project Engineer should consult them as an aid to determine where unsuitable material may be anticipated and to help plan the necessary excavation utilization.

203.04 — Classification

All excavation will be classified and the ordinary classifications are as follows:

A. Unclassified Excavation. This classification includes all materials encountered regardless of their nature or the manner in which they must be removed.

B. Classified Excavation.

1. Common excavation is that excavation not classified as rock or otherwise and includes all other materials encountered regardless of their nature.
2. Rock excavation includes all masses of material, which cannot be removed without blasting or ripping, and all detached stones or boulders having a volume of 2 cubic yards or more.
3. Borrow excavation is that material coming from approved sources normally outside the limits of the project. Unless otherwise provided in the contract, the Contractor is responsible for providing borrow sources at his/her own expense.
4. There may be an additional category of "unsuitable excavation" used to provide the removal and disposal of saturated or unsaturated mixtures of soil and organic matter not suitable for foundation material regardless of moisture content.
5. Precise records must be maintained on a project which has two or more classifications of excavation differentiated as separate pay items. Changes from one classification to the other must be noted and sufficient measurements taken to establish the correct pay quantities for each category.

203.05 — Unsuitable Materials

It is essential that the best possible use be made of the soil materials which are available for construction. There will be soils encountered that are unstable in the natural state because of excessive moisture content. However, many of these soils will respond to drainage improvement and/or mechanical manipulation, or combining with other materials, to reduce this moisture content and render them suitable for use. Further, many soils are unsuitable because of their natural composition and can cause instability in embankment or have some other detrimental effect. These should be wasted, or used to flatten slopes, unless an economical method of treatment is available to make them satisfactory.

203.06 — Salvaging and Placing Topsoil

In localities where the excavated material in the deeper layers of the soil is not conducive to good plant growth, or will result in a slow rate of growth, it is a usual practice to strip off the upper surface of roadway cut areas, embankment foundation areas and, in some instances, borrow sources. This material is then stockpiled (normally) and used later to cover completed cut slopes, embankment slopes, and other disturbed areas where re-vegetation is desirable.

Assistance should be given the Contractor in the selection of suitable stockpile areas that will minimize the haul involved and not interfere with or delay other construction operations during the progress of the work. During the stripping operation, more than just casual inspection should be given this operation. Although at this point there may be no measurements required for pay purposes, the quantity being stockpiled, and the manner in which it is excavated, is important. Excavating too deep will easily yield the estimate quantity, but the quality of the topsoil will be reduced. Shallow stripping will waste topsoil and the estimated yield will suffer. An estimated total of stockpiled material (based on load count, etc.) should be maintained to be assured that enough topsoil is stockpiled before excavation for embankment construction is begun.

The topsoil should be lightly incorporated with the existing subsoil for bonding to prevent topsoil slippage and sloughing. Slopes should be scarified or loosened to a depth of at least two inches. Roots and/or stones (generally larger than four inches) that turn up should be removed.

Equipment used in the placing and preparation of the topsoil should be at the Contractor's discretion provided satisfactory results are obtained. Normally such equipment will consist of farm tractors, plow and disc attachments, and cultipackers or rollers to break up dirt clods.

Stockpiles, if to be left in place for an extended period, should be mulched or seeded to prevent erosion.

203.07 — Surplus Material

It may be determined during the plan development stage of a project that the required excavation will yield more material than the proposed embankment construction will require. This surplus material, normally classified as waste, may be generated during embankment construction in approximately the amount anticipated or to a greater or lesser extent, dependent on the accuracy of the original shrinkage factor. The Contractor and Project Engineer should stay aware of these quantities and, as the work progresses, checks should be made to ascertain if shrinkage and/or swell factors are sufficiently accurate. These factors, when determined during plan development, are usually based on experience with the material involved and/or tests. However, field checks are recommended to substantiate these factors. By calculating the amount of excavation removed from a cut, whether partially or wholly complete, and comparing this amount with the quantity of embankment constructed with this material, a reasonable check can be made. Relatively large amounts of excavation/embankment increase the accuracy of this check. It is also more meaningful if the method of measurement and calculation is the same as that to be used in establishing the final quantities. If at all possible the embankments should be completed before any material is wasted. However, if this is not practical, the calculations indicated previously should still be made to check the validity of the shrinkage and/or swell factor in use to ensure that the remaining excavation is sufficient to complete the embankment item.

203.08 — Blasting

Blasting operations should always be conducted under competent, careful supervision to prevent injury to persons or damage to adjacent property. At the same time this operation should produce specification material and leave the "cut" area at the correct line and grade. The spacing and depth of drill hole, the explosive type and loading sequence, the method of firing, and kindred matters are conditions to be satisfied by the blasting experts. Each of these is dependent on local rock conditions and the production equipment to be used. Good blasting techniques can materially lower the Contractor's cost while producing specification material under relatively safe conditions. Poor blasting by unqualified personnel is not only dangerous but it can cause excess overbreakage or fail to shatter the material enough for economical handling and placement. Consequently, this latter material will require additional drilling and shooting.

The spacing and strength of the blasting charges should be such that the actual blast is as light as possible consistent with thorough rock breakage and reduction in size for economical hauling and placement. Overshooting is very undesirable. It can produce rock fractures beyond the intended line and grade that result in jagged slopes, uneven ditch grades, and possible future slides in the slope areas. Special care should be used in the blasting of material immediately adjacent to proposed slopes or ditch lines. In deep cuts through difficult material where slope smoothness is hard to obtain, it can be advantageous to drill slope holes parallel to the proposed slope. In these "sloped" holes the charge is distributed along the length of the hole, instead of placing all of it at the bottom, and then exploded simultaneously. This "pre-splitting" is not intended to loosen and break up a great amount of rock, but to shear a reasonably smooth plane along the proposed slope face. The rest of the rock in the cut is then shot and removed in a manner best suited to the material involved.

The Contractor should be required to perform all roadway blasting necessary at structure sites prior to beginning structure construction to avoid undetected damage to the structure components from blasting operations. When blasting near developed areas, a joint Department-Contractor inspection should be made of surrounding properties prior to blasting to assist in verifying any damage claims which may arise from the blasting. A camera should be used during this inspection.

203.09 — Transition Points

Transition points are defined as the points of change from cuts to fills. Many times roughness in the base and surface courses of a roadbed appears at the junction of sizable excavation and embankment sections. In all probability this is the result of a lack of attention to the importance of embankment foundation preparation in these critical areas. This situation is comparable to the junction of a side hill excavation and its laterally adjacent embankment. Under these conditions, "benches" are cut into the existing ground slope as each succeeding layer of embankment is placed. This bench area, compacted with this layer of embankment, is then covered with the next lift of loose embankment material. Then, as a result of the manipulation and compaction that follows, this embankment lift is "keyed" into the adjacent natural ground. This procedure, with the "benches" cut at an angle to the road centerline that parallels the ground slope, can be advantageous in reducing subsurface movement where these transition points occur. Consideration should also be given to proper drainage of these benched areas. In many instances, undercutting of the transition points will provide a better foundation for the base and pavement structure and should be routinely considered.

Embankment (General)

203.10 — Preparation of Embankment Foundation

Prior to beginning the construction of an embankment, the area to receive the embankment and serve as its foundation should be carefully inspected. All locations of questionable supporting capacity should be given special attention. The presence of soft or very wet conditions may suggest the need for removal of certain unsuitable materials, installation of underdrain facilities to remove spring or seepage water, or merely the need for aeration and drying the materials that have become saturated due to poor surface drainage. Conditions of this type should always be called to the attention of the Project Engineer for decisions concerning appropriate action.

The plans and specifications should be carefully followed in preparing the foundations for embankments which are to be constructed through swamp, marsh, and old lake bed areas. Special construction methods are usually required in these cases and are generally worked out during the design stages for a particular project.

In constructing embankments on steep slopes, particular attention should be given to obtaining a good interlock between the sloping foundation and the new embankment. Under the steepest slope conditions, all vegetation should be removed to prevent the formation of slippage planes. Proper interlock between the surfaces can usually be provided by plowing or ''benching'' the old ground surface. Hard surface areas should be plowed, terraced, or benched prior to the placement of any embankment on them. This method of keying the new embankment to the foundation surface should also be employed when widening and/or raising the grade of old embankments.

203.11 — Earth Embankment

A. General. The importance of uniformity in embankment construction cannot be overemphasized. Construction methods which ensure, to the extent economically feasible, uniformity of material, layer thickness, moisture content, and compactive effort are vital in the accomplishment of proper embankment construction. The amount of inspection needed to secure proper construction will depend primarily on the size of the project, the nature of the soils, the rate at which material is being placed, and the governing specifications.

B. Uniformity of Material. The proper breakdown of clods and blending of materials is very essential in obtaining proper embankment consolidation. The use of disc plows, blade graders, or similar equipment ordinarily will accomplish the desired result in most soils; however, the use of mechanical pulverizers may expedite the work of processing some of the very plastic clays for compaction. When it is necessary to place unlike materials in the same embankment layer, care should be taken to use equipment and methods which will provide a satisfactory blend of these materials. The end result of the blending should be a mixture which can be uniformly compacted after moisture has been applied using a uniform rate of application. When feasible, poorer materials should be used in the lower portion of the embankment.

C. Hauling, Spreading, and Shaping. Prior to beginning the construction of an embankment, the location of the construction should be clearly outlined by construction stakes. Any staking that is done by the Contractor should be checked by the Department's inspection forces. Stakes which are disturbed or destroyed should be reset immediately. The use of this system of checks will make it unnecessary to correct deficiencies in embankment slopes or widths after construction of the embankment has advanced to the point where corrective measures are difficult and costly.

Washes, holes, and other low areas in the embankment foundations should be filled and compacted prior to beginning the construction of the first lift of embankment. The first lift should begin in the low areas with the intent of eventually providing an embankment layer approximately parallel to the finished grade. As the earth materials are being dumped and spread, large roots and other perishable materials must be removed and disposed of outside the limits of the work. Good drainage must be maintained throughout the construction period. Frequent checks should be made to see that the work conforms to the proper cross section, all drainage facilities are open, and that the embankment site is in the best condition possible to provide maximum runoff.

The selection of the equipment used to haul the embankment materials is usually the Contractor's option; however, the Contractor should not be permitted to operate the equipment in such manner that it will be harmful to any section of the completed roadway. The soil may be spread with the hauling equipment or it may be spread by the use of blade graders, bulldozers, or other acceptable equipment. Construction hauling over expansive type clays should not be confined to the same path but should be spread out over the entire width of the embankment to avoid ruts or overcompaction.

D. Layer Thickness. More uniformity of compaction is ordinarily obtained when earth embankment layers are relatively thin. Past experience indicates that uniformity of compaction can be obtained by constructing layers ten (10) inches or less in loose depth.

E. Moisture Content. The success of compaction operations is dependent to a large extent on proper moisture control. If the proper amount of moisture is uniformly distributed throughout the embankment layer, rarely will there be any difficulty in obtaining satisfactory compaction, provided the thickness of the layer does not exceed the capabilities of the roller being used. The common tendency to construct earth embankments at moisture contents on the dry wide of "optimum" makes the task of securing uniform moisture distribution and satisfactory compaction more difficult. Usually, it is better to begin compaction with the moisture content slightly high; however, in humid areas it may be better to begin with the moisture content at optimum.

The mixing and blending of soils and water should be thorough. Large clods and lumps must be broken down to ensure a uniformly moist condition. Whenever it is necessary to blend moisture into very plastic clays, heavy plowing and turning of the soil will usually do the job satisfactorily; however, in certain extreme cases some type of mechanical mixer may expedite the work.

When adding water to a layer of material, care must be taken to avoid overlapping or gapping between successive passes of the water distribution equipment. Wet or dry streaks are undesirable and should be avoided. The equipment operator should be instructed to begin applying water on one side of the embankment and work progressively across the fill to the other side to avoid having wet or dry streaks in the center of the embankment. The importance of making several light applications rather than one heavy application of water should be emphasized.

203.12 — Earth Embankment Compaction

A. Stability of Embankment. Embankment solids must be compacted to densities which will provide adequate stability under the conditions to which the finished highway will be subjected. Experience and the results of considerable research have provided a means for determining the desired in-place density for the various soil types. Ordinarily, nonplastic and moderately plastic soils are compacted to the highest practical density at or near the optimum moisture content. If it is not feasible to confine expansive soils to the lower portions of large embankments, control of over-compaction as well as under-compaction may be justified. With these soils, high densities obtained at low moisture contents are not desirable since swelling, loss of stability, and roughening of the finished pavement may occur after construction is complete. With certain soil types such as the more plastic clays and silty clays with high moisture content, soil stability can be enhanced by the addition of lime or cement to the soil. Laboratory testing for compatibility and rate determination is necessary when this alternate is considered.

B. Density Tests for Control of Compaction. When in-place tests are to be used for controlling the compaction of embankment, the types of equipment usually are not specified. The Contractor has the option of using any method or any equipment desired. Samples of the various soil types to be compacted should be obtained and the required moisture and density relationships determined in the laboratory prior to starting compaction operations. To delay this work until after construction begins may retard the Contractor's progress or make the compaction control procedure ineffective. However, during construction additional tests should be made on any soils which were overlooked during the preliminary sampling operations. Soil profiles which were made during the design stages for the project will usually prove to be helpful in determining the required number of samples for test.

In-place density tests for checking compaction should be made using the specified equipment and in accordance with the specified test procedures. Procedures which simplify the problem of soil identification should be devised and used whenever possible. In addition to providing laboratory density values and optimum moisture contents for each of the soil types, the report to the field should include all information which can be used in soil identification procedures. Information relating to soil plasticity, gradation, color, and texture for each soil type will usually prove to be very helpful in selecting the appropriate density value to use in checking the in-place construction. If identification of the soil sample for density is impractical, density control should not be used.

In using density tests to determine the adequacy of compaction, recognition should be given to causes of variations in test results which are to be expected in using this method of compaction control. Exact identification or exact representative samples are seldom possible. Mixed soil conditions and use of soils of varying rock content are specific factors which frequently are the cause of difficulty in the interpretation of test results. When it is not possible to properly perform meaningful tests due to high rock content or for other reasons, explanations as to why the tests were not made should be entered on the form provided for recording the test results. In cases where materials are predominantly rock and tests are not feasible, proof rolling may be justified.

C. Control of Compaction by Ordinary Methods. If the specifications require that certain equipment be used, it must be inspected and approved prior to its use. This information is to be made a part of the job records. Particular attention should be given toward maintaining (a) proper loose layer thickness, (b) proper moisture content, and (c) adequate coverage with the specified rollers. If the specified roller is suitable for the materials being used and is of the proper weight, embankments constructed in relatively thin layers at optimum moisture content will normally possess the required stability, and neither settlement nor swellage due to improper consolidation will be a serious problem. Proof rolling may be deemed advisable on some projects to determine if the equipment and methods being used for obtaining compaction are satisfactory.

203.13 — Rock Embankments

A. Placing. The construction methods for placing embankment material consisting principally of rock usually will depend on the size of the rocks and the amount of rock present. Ordinarily, rock embankments are constructed in layers extending over the full width of the roadway, with the layer thickness conforming to the requirements of the governing specifications. By exercising skill in handling, the coarse and fine materials can be usually be distributed so that the interstices in the various size stones will be filled with small stones and earth to make the embankment as dense and compact as possible. When material is placed in the fill by the end-dumping method, this work should be accomplished by dumping on top of the layer being built and shoving ahead into place with a bulldozer. Allowing material to roll into place by dumping over the edge of completed work should not be permitted. Oversize rocks not suitable for placement in a layer should be broken down to the proper dimensions or moved to a portion of the fill where large rocks can be satisfactorily used. Caution should be used to avoid placing rock embankments in conflict with the future installation of culverts, guardrail posts, or foundations for lights and signs.

B. Compaction. When rock is present in the embankment material in considerable proportions, moisture control procedures are of little benefit and density tests for checking compaction are not considered feasible. Heavy rollers frequently prove to be advantageous, particularly when combined with vibratory rollers.

203.14 — Finishing Earthwork

A. Roadbed and/or Subgrade. Before beginning the work of preparing the roadbed and/or subgrade for placement of the pavement, the entire roadway should be carefully checked to see that the earthwork conforms to be required cross section and grade within the accuracy or tolerances established by the specifications. Any unstable or soft areas should be corrected by aeration or by removal and replacement of unsatisfactory materials. Areas which appear to be excessively wet should be investigated for seeps and other sources of water. If such conditions are found, the Project Engineer should take appropriate action for correction of the conditions.

Unless otherwise specified, excavation in solid rock should extend to at least 6 inches below subgrade. These cuts below grade should be backfilled with selected material conforming to specified requirements. Isolated outcrops of ledge rock or portions of large boulders which are exposed should be removed to a depth of not less than 6 inches below the subgrade elevation. Low areas caused by such removal should be backfilled with suitable material. Following the removal and replacement of unsatisfactory subgrade materials, the subgrade should be scarified and all deleterious material removed. Large stones shall be broken down or removed from the top portion of the subgrade in order not to interfere with subsequent fine grading operations. The loose material should be brought to the proper moisture content and compacted. The stability of the roadbed and/or subgrade should be maintained as existing conditions may require until the subbase and/or pavement is placed.

B. Slopes, Shoulders, and Ditches. When feasible, the finished earthwork should have a pleasing contour and be reasonable smooth in all respects. A slight rounding effect at the tops and bottoms of slopes will help give a pleasing appearance and reduce erosion. Whenever excess excavation is available and conditions permit, a rounding out at the toes to fill slopes is desirable. This practice will usually increase the stability of the embankment and also reduce erosion. Transitions in slope and warping to fit special or unusual features within the roadway should be gradual. Abrupt changes of slope at the intersections of cuts and fills should be avoided whenever possible.

Economic considerations usually make it impractical to trim rock slopes to the exact cross section; however, any dangerous or objectionable appearing projections or partially embedded boulders should be removed. The finished slope should have a uniform appearance.

The practice of constructing earthwork to reasonable neat and uniform lines is desirable for all phases of the earthwork construction, including channels, roadway ditches, and other such features. Shoulder construction generally should be kept current with other phases of the roadbed construction.

The work of finishing earthwork can be held to a minimum and better construction will result if every reasonable effort is made during excavation of cuts and construction of embankments to maintain slopes, shoulders, and ditches as this work progresses.

203.15 — Hydraulic Fill

There will be occasions when it may be desirable to obtain embankment material from sources under or adjacent to water. When the material selected for embankment use is beyond the reach of dragline equipment and located under swampy ground, river, or lake bed, excavation and movement of the material can be done efficiently by hydraulic dredging. Even if the material is not covered by, but is adjacent to, water, this equipment may prove very competitive with other types of excavation equipment. The dredge pump draws water and suspended material through the suction pipe and then forces this mixture through a line of pipe to the point of discharge. Some materials can be excavated by suction alone, but the tougher, more cohesive materials require agitation of cutting to loosen and stir them up prior to pickup by the suction intake.

Frequently the embankment foundation area contains an overburden of muck or other unsuitable material which must be removed prior to placement of the excavated material. If job site conditions permit, this may be handled by the same equipment that will be used to excavate and place the embankment material. Suitable disposal areas must be provided for all unsuitable material removed from the embankment and/or borrow source areas.

Selection of material from within the area to be excavated must be given the same careful attention as if it were from a surface borrow source. All unsuitable material should be removed from the entire area before any embankment material is excavated. During production, the material being discharged on the embankment area should be observed for contamination due to an abnormal amount of unsuitable material. This material, such as clay balls in sand, has a tendency to settle out in the vicinity of the discharge and form a pocket of undesirable material which can produce instability.

The embankment material is usually pumped through the pipe system directly to the embankment site where it is shaped and compacted by a bulldozer and/or other suitable equipment. In other cases, it may be stockpiled at an intermediate site and then hauled to the embankment site in conventional haul equipment. The governing specifications will then dictate the manner of spreading, manipulation, and compaction.

203.16 — Records

Particular attention must be given to the keeping of accurate, up-to-date records of all the work. Some of the more important items which should be documented and made a part of the project records are:

1. Locations of actual balance points and notes concerning cross-haul that may have occurred, including reasons for the cross-hauling and the quantities involved.
2. Measurements and notes made to substantiate the classification and quantity of the various materials encountered in the excavation.

3. Records of dimensions, weights, and calibrations which may be required by the specifications for water distribution and rolling equipment.
4. Measurements of excavation below grade necessary for the removal of unstable or other undesirable material.
5. Daily record of events, including limits in which work was performed during the day, type, and number of pieces of equipment used, number and classifications of labor used, and notes regarding discussions of an unusual nature with the Contractor's representative and which may lead to the issuance of specific recommendations or instructions to the Contractor.
6. A complete record of all density tests made to determine the acceptability of embankment, with remarks included which explain the corrective action taken at locations where failing tests results were obtained and reasons for not making the required tests at other locations. The results of proof rolling should also be recorded.
7. Complete record of equipment, labor, and materials used in an operation that is, or may later be determined to be extra work.
8. Notes regarding damage to private property caused by the Contractor's equipment and/or operations.
9. Record of final disposition of salvable materials.
10. Weather conditions and their effect on quality and workability of earthwork materials.

203.17 — Measurement for Payment

Project records must be kept in such a manner that all partial progress and final payments can be easily and clearly supported by recorded data. This data should include, when the work was completed, calculations to support the quantity allowed, and should be initialed or signed as required by the person who made the calculations and measurements. It is recommended that a continuous record be kept, immediately subsequent to their completion, of the items that are paid for as each linear foot, square yard, etc., and this record must include the measurements made to substantiate the quantity allowed.

Section 204 — Subgrade Preparation

Subgrade Preparation consists of scarifying the top six inches of the subgrade, mixing until uniform in color and texture, compacting to a higher density than normal embankment layers at a specified moisture content, and shaping to the required typical section within specification tolerances.

Prior to scarifying the existing material, the subgrade should be checked to ensure that it is generally within the same surface requirements as will be required for the finished subgrade surface. After the top six inch layer has been processed, it should be bladed and shaped so that it is within 1/2 inch of the prescribed elevation at any point.

Section 205 — Prewatering of Excavation Areas

If additional moisture is required in excavated material for proper compaction, water may be added at the source of excavation or borrow, or on the roadbed. Prewetting of the material in the excavation area or borrow source may be used if it is practical. This watering may be done, if feasible because of soil type and water availability, by sprinkling, flooding, or irrigating the area. This is done not only to satisfy the need for additional moisture in the soil, but also because a more uniform distribution of moisture is usually obtained. In addition, machine manipulation of the soil on the roadbed is reduced by that amount that would have been used to incorporate added water into the soil.

Section 206 — Overhaul

Overhaul is the hauling of excavated material beyond the free-haul distance in moving this material from the point of excavation to embankment or other disposal area. Normally, it is more economical to haul roadway excavation quite some distance to construct an embankment rather than to waste this material and use borrow excavation. However, there are times during construction when a planned sequence of operations must be altered because of unforeseen delays or for reasons of economy. Such a change could require the substitution of borrow for long-haul excavation. If this is the case, changes in the mass diagram and overhaul quantities will occur. Therefore, accurate records of dirt movement must be maintained at all times so that a true total haul determination can be made and supported later. These records should reflect the balance points as were actually used for the grading operation, any changes ordered by the Project Engineer, and any unordered cross-haul by the Contractor. Unless these balance points and cross-haul notes reflect the true position of the material in the roadway, there can be no accurate swell and shrinkage factors or overhaul quantities calculated.

Section 207 — Structural Excavation for Conduits and Minor Structures

The inspector should refer to Section 207 of the AASHTO Guide Specifications.

Section 208 — Erosion Control

It is essential that every possible measure is taken to control erosion and sediment pollution during highway construction. Materials and devices such as berms, dikes, dams, sediment basins, silt fences, netting, gravel, mulches, grasses, slope drains, ditches, channels, riprap, and fiber mats may be utilized as temporary control measures until permanent controls are installed and become effective.

Where temporary erosion control features are not included in the plans, it is expected that the Project Engineer and the Contractor will anticipate possible problems and provide timely and adequate controls to prevent or at least minimize adverse effects upon the environment.

The most effective erosion and siltation control that can be exercised on any highway construction project is early treatment of the slopes—both cut and fill. Early treatment means treating cut slopes as excavation progresses and fill slopes as embankment construction proceeds. Slope treatment varies from State to State, but generally consists of seeding and mulching.

All slopes shall be stabilized by mulching, seeding, or otherwise protected as the work progresses. All damaged slopes should be repaired as soon as possible. The Project Engineer should limit the surface area of erodible material exposed if the Contractor fails to sufficiently protect the slopes to prevent pollution.

The Contractor should at all times have on hand the necessary materials and equipment to provide for early slope stabilization and corrective measures to damaged slopes.

The Contractor should operate all equipment and perform all construction operations so as to minimize pollution.

The Project Engineer must continuously assure that all installed devices, both permanent and temporary, are maintained in an appropriate condition such that they can perform their intended function. This may require clean out of deposited material, replacement of component parts, and rebuilding as may be necessary.

DIVISION 300 — BASE COURSES

Section 301 — Plant Mix Asphaltic Base Course

Instructions to Plant Inspectors and Roadway Inspectors are contained under Section 401 of this manual.

Special points relating to base course construction are:

1. *Consecutive course placement.* Succeeding courses should not be placed until the previously placed course has cooled to the extent that it will not be distorted by equipment.
2. *Compaction of shoulder.* Two types of special equipment are commonly used in trench widening—trench rollers and vibratory compactors. Either must generally be supported from adjacent pavement and must be adjustable to bear fully on the course to be compacted.

Section 302 — Road Mix Asphaltic Base Course

Instructions contained in Section 403—Road Mix Asphalt Pavement will apply to this work.

Section 303 — Reserved

Section 304 — Aggregate Base Course

304.01 — Description

This course is placed on prepared surfaces to (1) distribute the wheel loads transmitted to the subbase, (2) provide a free draining material, and (3) to provide a non-frost susceptible material on which to support surface courses. It is placed in various depth and width combinations, depending on the particular location and existing soil conditions.

Usually the depths are designated on the plans and cross sections but occasionally it may be necessary to modify the depths because of unforeseen conditions found after construction has started. It will be the duty of the Inspector to call to the attention of the Project Engineer all conditions which appear to need changes in base depths.

304.02 — Materials

Typical materials used to construct base courses are sand, gravel, crushed stone, slag, sea shell, or a combination of these or other readily available granular-type material. They may be either pit run or processed, depending on the material available and the specification requirements.

The gradation requirements are outlined in the specifications. Since the requirements are part of the contract, it is imperative that the Contractor supply material which will comply. No changes can be made at the project level. If justified and supported by approved change orders, changes from the specified gradations may be allowed. Any changes should indicate a benefit to the State. Such changes should not adversely affect the integrity of the design.

The Inspector must be thoroughly familiar with the plans, cross sections, specifications, and the contents of this manual in order to be knowledgeable in all the construction requirements. The Inspector must be assured that all preliminary tests have been made and that the proposed material will be acceptable when placed. Frequent discussions with the Project Engineer will no doubt be necessary, particularly when base operations are first started. This is a crucial period since many trials are sometimes necessary before the Contractor can produce consistently acceptable aggregate. Regardless of the difficulties encountered no base should be accepted which does not meet the requirements. If substandard material is inadvertently placed it should be removed or upgraded. The Contractor should not, however, be allowed to knowingly haul substandard material onto the work with the intention of later upgrading the aggregate in place.

304.03 — Sampling and Testing

Sampling and testing of the proposed aggregates should be done in accordance with AASHTO T2 or other specified methods. As the basing operations proceed, testing should be done on a continuing basis. Maximum density and optimum moisture should be determined in accordance with AASHTO T99 Method D or T180 Method D.

304.04 — Placing

Prior to placing base, the surface upon which the material is to be placed should be carefully examined. Soft spots, ruts, and grade deficiencies should be corrected by removing, replacing material, or regrading where necessary. The top of the subbase should be shaped to the required grade and cross section and smoothed to provide drainage. The Contractor should plan his hauling so that truck traffic is distributed over the entire width of the surface.

The loose aggregate should be placed a little in excess of the specified maximum thickness to allow for compaction. The aggregate should not be dumped in piles but should be spread either by the truck traveling as it empties or by use of spreaders. If spreaders are used, the box should be kept full. Precautions should be taken to avoid segregation; that is, the large stone separating from the fine portions. If segregation does occur, the Contractor should be required to remix the aggregate by blading, rototilling, harrowing, or by other method.

The Inspector should be constantly alert during placement to detect changes in the appearance of the aggregates, particularly on pit run material, so that tests may be made immediately on the changed material. Oftentimes excessive silt, clay, or organic matter inadvertently become mixed with the base material and should be removed. An excess amount of these materials is very detrimental to the base since they (1) increase the susceptibility of the material to the action of frost, and (2) reduce the strength of the base when it is saturated with water. Constant vigilance accompanied by testing will help assure that detrimental materials are not used. Also, oversized rocks should be removed since uniform compaction is difficult to obtain if they are present. Alternate freezing and thawing may tend to move the oversized rocks vertically, ultimately producing bumps in the finished road surface.

304.05 — Compaction

Following closely behind the placing operation should be the compaction. There are basically three factors which influence the compaction: (1) moisture content, (2) gradation, and (3) compactive effort.

The moisture content at which compaction can be most easily obtained, with a stated amount of compactive effort, is referred to as the Optimum Moisture. Laboratory tests will reveal the moisture content which constitutes optimum moisture. The in-place density should be determined in accordance with AASHTO T238, T205, or other specified density test. If the natural moisture content of the available material is found to be low, compared to the Optimum Moisture, as it may easily be for granular-type material, water or calcium chloride, or both, should be added as indicated by plans and specifications.

The water and calcium chloride may be introduced by (1) a passing of the materials through a stationary plant, mixing the additives thoroughly, or (2) they may be added to windrows on the road and mixed by blading back and forth or rototilling, or (3) they may be added by using a traveling plant mixer. In any case, thorough and uniform distribution of the additives must be obtained throughout the material. Moisture added to the surface aids only the material near the surface. The material at the bottom of the layer does not receive the needed moisture and therefore ultimate compaction will be less at that level. Water must be added to the surface to replace that lost by evaporation during processing.

Machinery used to apply the compactive effort may be any one or a combination of the following:

1. Pneumatic-tired rollers
2. Vibratory rollers
3. Steel-wheeled rollers
4. Pan-type vibrating compactors

The Contractor will usually be free to choose the type of equipment most adaptable to the material and work, subject to whatever requirements are specified in the contract.

Usually the specifications will state how much compaction will be required, unless standard or ordinary compaction is permitted as per Sec. 203. This is normally expressed as a percentage of the maximum density achieved with the "Standard Proctor" (AASHTO T99) or "Modified Proctor" (AASHTO T180) tests. The in-place density should be determined in accordance with AASHTO T238, T205, or other specified density test.

Once the required moisture content and compaction is achieved it must be maintained. This can be accomplished by the continuous addition of small amounts of water to replace moisture lost by drying of the surface.

304.06 — Checking Grade

The final duty of the Inspector as regards inspection of the aggregate base course construction is checking the grade and shape of the finished aggregate surface. This can be done in one of several ways: (1) hand levels, (2) string lines, (3) cross sections. Grades should be checked every 50 feet on tangents and curves less than 16 degrees and every 25 feet on curves greater than 16 degrees.

A. Hand Level Method — Grade Check. When using the hand level, the inspector sights through the tube at a known grade, usually marked on a grade stake, and measures the distance vertically from the level sight to the ground elevation. This distance is compared with the calculated distance as obtained from typical cross sections and grade elevations for the particular location being checked. Hand levels should not be used for sights exceeding a distance of 50 feet. Grades placed on stakes to be used for checking grade should be established with a transit (or level) and a level rod.

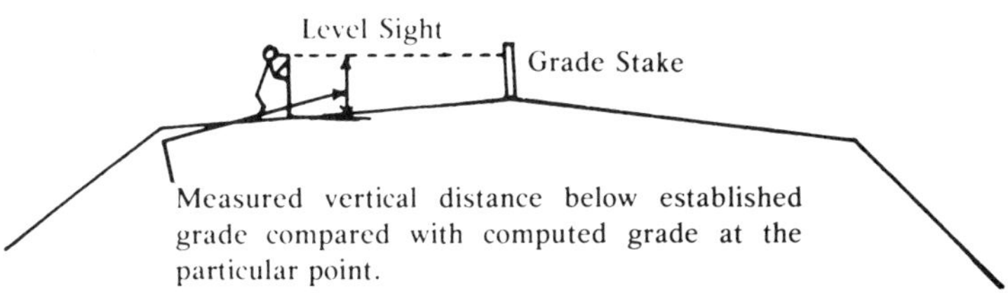

B. String Line Method — Grade Check. The string line method is applicable when the distances are not so great (about 25 feet or less) that sag in the string line is excessive. After placing level grades on stakes across the road, a string is stretched between stakes and the vertical distance at various points measured and compared to the computed dimensions.

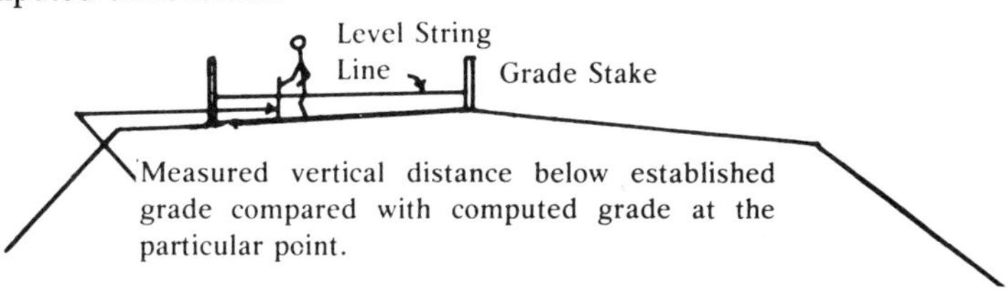

C. Cross-Section Method — Grade Check. The cross-section method is adaptable where comparisons with the required section are difficult because of the complicated shapes. These sections are taken with transit (or level), recorded in the normal manner: listing station, offset, and rod reading. The cross sections are plotted and compared with the required shape. This method is time-consuming and should be resorted to only in unusual cases.

The finished surface of bases shall also be shaped within reasonable conformity to the design cross section. The Inspector shall check the base for compliance at intervals of not more than 50 feet and shall note in the diary that the checks were made. A complete record of the measurements shall be made for a minimum of one section in each 500 feet increment.

304.07 — Measurement and Payment

Payment is made by the unit price per ton, square yard, or cubic yard. Measurements should be made as outlined in the plant mix pavements section of this manual regarding scales and delivery tickets.

Water added to the materials, unless otherwise specified in the bid documents, will be paid by the 1,000 gal. unit (M.G.) at the contract price and should be supported by delivery tickets. The project record must contain a record of the calibration for tanks or distributors or water meter checks and any necessary adjustments. Calcium chloride will be paid for by the ton. If sacked calcium chloride is used, the net weight printed on the sack is acceptable for measurement.

304.08 — Records

Recordkeeping is a very important part of the Inspector's work. Observations, measurements, and directions are the basis for justifying all parts of the work. Because much of the Inspector's work is covered by subsequent construction, the results of the work cannot be readily reviewed later. Written reports and records of the observations and measurements are usually the only remaining evidence that the work was performed correctly and that the State received the complete benefits of the Contractor's work paid for. Emphasis should be placed on recording all portions of the work daily as it is performed.

The procedure to be used to document the aggregate base course is as follows:

1. Each grade Inspector shall make a written statement, in a bound project field book, each time a series of grade checks is made. Statements should reflect both the work and the findings. They might be of the following nature.
   - "Checked grade Sta. 100 + 00 – 105 + 00; found it to be within limits specified; more work needed."
   - "Checked grade Sta. 110 + 00 – 115 + 50; found it to be within limits specified; gave contractor go ahead."
   - "Rechecked grade Sta. 100 + 00 = 105 + 00; found it to be within tolerance permitted by plans and specs.; told Contractor to go ahead."
   - Statements should be dated and signed.

2. All notes and computations are to be retained and made a part of the project records regardless of their form or condition.

Other records necessary to document this work include—

1. Gradation test reports to show the quality of the material used.
2. Compaction tests to show that the required compaction was achieved. This should also include tests for moisture content.
3. Measurements for final payments, such as cross sections, delivery slips, or in-place measurements are required.

Section 305 — Subgrade Modification

305.01 — Description

This item provides for modifying the subgrade layer with materials obtained from the ditches and shoulder slopes and/or with aggregates and additives. Aggregates and additives which may be included in the subgrade are water, calcium chloride, and sodium chloride as shown on the plans.

The addition of calcium chloride to the subgrade material is done for the purpose of improving the physical characteristics of the material. One of the benefits of adding calcium chloride is the higher densities which can be obtained and the improved stability which results. Other benefits of calcium chloride include less compactive effort requirements to achieve density, optimum moisture control by inhibiting water evaporation, a reduction in the amount of fines required in the subgrade for cohesion and improved surface uniformity.

The addition of sodium chloride is most beneficial in climates affected by frost and in soils or aggregates that have sufficient fine-grained material to react. Soils high in organic matter do not react favorably with sodium chloride. Benefits of sodium chloride stabilization include lowering the evaporation rate of water, less compactive effort and better cohesion much like calcium chloride. Inclusion of salt generally permits the use of more readily available materials which have more fines than would be desirable otherwise.

The quantity of additives and water necessary will be determined by tests of samples of the subgrade to be treated. The rates will be specified on the plans or as directed by the Project Engineer based on the results of the tests. The depth and width of the treatment will be designated on the plans.

If material is utilized from the shoulders and ditches, these areas shall be realigned, cleaned, and reshaped in conformance with the typical sections shown on the plans. This work shall be performed after all scarification, additive mixing, and compaction have been completed.

305.02 — Materials

Additives used will be dry powder, pellets or mixed in a slurry. It will usually be delivered to the project in bulk.

305.03 — Sampling and Testing

Sampling and testing of the soils to determine the percentage of additives and water must be done.

305.04 — Equipment

The Inspector should inspect all equipment specified or proposed to be used, for conformity to the governing specifications.

305.05 — Modifying Subgrade with Existing Material

Material incorporated from the shoulder slopes and ditches must be spread uniformly over the roadbed-surface and all weeds, sod, grass, roots or other objectionable material must be removed. The roadbed will be scarified and all material thoroughly mixed and pulverized until not more than 5 percent exclusive of gravel or stone is retained on a 2-inch sieve. If additional material is required to obtain the depth as specified in the plans, it shall be obtained from approved locations and blended with the existing material in a uniform windrow.

Sufficient water will be added or removed during the mixing operations to provide the optimum moisture content, plus or minus 2 percentage points. The subgrade layer shall be compacted to a density not less than _____ percent (95 suggested) of maximum density.

Once completed, the surface shall be maintained until the placement of the base or surface course. Additional water shall be applied as needed to prevent checking or raveling.

305.06 — Mixing Operations, Compaction, Finishing and Curing

A. General. When materials are to be mixed and blended, equipment used shall be so designed, operated and controlled as to deliver a uniform mixture of all ingredients with a minimum number of passes of the machine.

Proper moisture content must be reached and maintained throughout the mixing and compacting process. Moisture should be held within 1-2% of optimum. Caution should be used in applying water. If applied too fast or too heavily, the water content may be greatly in excess of acceptable range, and time is lost while waiting for the excess to be absorbed or evaporate.

B. Motor Grader Mixing. When the motor grader method is used, the additives may be applied over the subgrade in many ways. If the aggregate to be stabilized is spread over the length and width of the road, a spreader with uniform discharge and ease of discharge rate control may be used. The more uniform the initial spreading of the additives, the less work will be necessary to obtain a uniform concentration by mixing operations.

The additives and aggregates are then moistened and thoroughly mixed with a motor grader, etc. If a motor grader is used, the material should be completely turned at least four times at a speed such that the material will roll on the mold board. The water content should be such that after mixing, the material when squeezed in the hand will remain in a rough cast when the pressure is released.

After the materials have been thoroughly mixed, they should be bladed into windrows on each shoulder making as even a distribution as possible.

Inspection and compaction of subgrade should be made at this time. During the mixing operations, care should be exercised to avoid cutting through the subgrade and thus increasing the amount of soil in relation to the amount of additive which has been added. Such a practice may also change the gradation of the material to be stabilized.

C. Compaction. The capacity that the roller is designed to compact should not be exceeded. This is usually equivalent to 4 to 5 inches of loose material. The recommended lift is usually 4 inches. The spreading should be done in such a way that the uniformity of the material and its optimum moisture content will be maintained.

If the road mix method is used, blade approximately three inches of windrowed material over the wetted subgrade or subbase, alternating from one shoulder to the other. Materials should be held at or near optimum moisture content. Rolling should be started immediately, starting at the shoulder and progressing to the center. This procedure is followed until all windrowed material is brought back.

Vibratory, pneumatic, or steel rollers should be used according to design.

The density of the compacted modified course should be measured as rolling proceeds and stopped at optimum. Over rolling can cause a rapid loss in density.

D. Finishing and Curing. Rolling and shaping of the layer, while maintaining crown, should be continued until all materials are compacted into a dense mat and the surface assumes a dry appearance. Steel wheel rollers are best adapted to finish rolling. The final lift should be a minimum of three inches. The surface should then be sprinkled until it is covered with free water and rolled in a float of free water. It should be rolled until sufficient mortar is brought to the surface to thoroughly bind and seal the aggregates.

The curing period depends to some degree on the weather. The road should be allowed to cure at least 15 days. Traffic can be permitted on the road almost immediately. As a matter of fact, in some areas of the country, the road is never closed, not even during the stabilization process.

Low traffic count roads may be used as open surfaces for a short period of time. The open stabilized surface will be found to dust much less than with unstabilized material. The surface will be quite hard and blading will only be possible after wetting by rain or sprinkler. As traffic increases, consideration should be given to covering them with a wearing surface.

305.07 — Measurement and Payment

Payment for this work is described in the specifications. Measurement of the additive aggregate and water, paid by the ton and M.G. (thousand gallons), respectively, must be supported by weight and delivery slips signed by the Inspector when the material is incorporated into the work.

For processing the subgrade, the station-to-station limits, square yards, or miles of the completed work should be recorded to determine the quantity or modified subgrade. Blading of ditches and shoulders; scarifying and pulverizing of existing roadbed; placing and mixing of materials on the road; disposal of unsuited material; finishing of the surface; and maintenance of the completed surface shall all be considered subsidiary to this item when that work is required on the plans.

305.08 — Records and Reports

Daily entries should be made in the diary book describing location of the work, instructions received from the Project Engineer, instructions given to the Contractor, unusual conditions, and other items of interest.

For density control projects, a tabulation should be kept of all the density tests made. This tabulation should include the following information: date of test, location of the test hole, (DA) density, estimated moisture content, actual field density, actual moisture content, and percent of (DA) density obtained. Failing test results should be referenced to subsequent tests and notations made as required to explain the action taken.

Section 306 — Reconditioning

This section of AASHTO *Construction Manual for Highway Construction* contains sufficient detail to guide the Inspector. Methods of checking quantities and records to be maintained are covered in other sections of this manual.

Section 307 — Lime-Treated Courses

307.01 — Description

The addition of hydrated lime to subgrade material is done for the purpose of improving the physical characteristics of the material. One of the most advantageous uses of lime is its ability to reduce the plasticity index of clay soils. Lime treatment of clay soils tends to reduce and minimize volume changes that usually take place in an untreated clay soil. Clay particles tend to agglomerate when lime is added and this improvement may be reflected by a change in the soil classification. Strength and durability of certain soils are improved with the addition of lime. The beneficial effects are generally attributed to complex chemical reactions that occur within the soil-lime-water mixture.

The amount of lime and water necessary will be determined by tests of samples of the soil to be treated. The rates will be specified on the plans or as directed by the Project Engineer based on the results of the tests. The depth and width of treatment will be designated on the plans.

307.02 — Materials

Lime used will be dry powder or mixed in a slurry. It will usually be delivered to the project in bulk. The Inspector should be sure that a Certificate of Compliance is available stating the material delivered complies with the specification requirements. The Inspector should be aware of the highly caustic and corrosive nature of quick lime, and should assure that appropriate preventive and protective measures are exercised by those working with this material.

307.03 — Sampling and Testing

Sampling and testing of the soils to determine the percentage of lime and water to be used should be done in accordance with AASHTO T86 and T87. Methods used to determine the various soils constants should be done as outlined in AASHTO T89 through T93 inclusive. From this information, plus whatever additional is needed, the laboratory will determine the percent of lime to be used.

307.04 — Preparation of Roadbed

The subgrade should be checked by the Inspector prior to placing lime to be sure the grade and shape are within reasonable conformity to the cross section required. Soft spots, ruts, and grade deficiencies should be corrected by removing, replacing, or regarding where necessary.

307.05 — Equipment

The Inspector should inspect all equipment which may include pulverizers, rototillers, traveling plant, water wagons, graders, rollers (pneumatic-tired and steel-wheeled), and all other equipment specified or proposed to be used, for conformity to the governing specifications.

307.06 — Application of Lime

Lime is usually delivered to the project in self-unloading truck transports. Each truck shipment must have the weight of lime certified on public scales, or the weight of lime must be determined on a set of scales furnished by the Contractor. It is essential that the weight of lime be known prior to its use.

If the Contractor proposes to mix the lime and water in windrows, the material must, after grading, be excavated to the required depths. If the Contractor elects to mix with rototilling type machinery, such removal will not be necessary. In either case, grading and rolling prior to lime application must be done in order to obtain the specified final grade and depth of treatment.

Unless otherwise specified, the Contractor has the option of placing the lime on the road either by the "Dry" or the "Slurry" method.

Lime must be distributed with a reasonable degree of uniformity. Dry lime must not be spread when the wind will cause an excessive loss, or when blowing lime becomes objectionable. In distributing lime, a uniform velocity of the vehicle must be maintained and care exercised to secure lapping of the longitudinal strips.

Frequent checks on the rate of application must be made. The actual area of spread for each shipment should be compared with the desired area of spread, as determined from the specified rate, and such information made a part of the job records.

When the slurry method of application is used, lime and water are mixed and applied through spray bars from tank trucks equipped with agitating equipment to keep the lime in suspension. The procedure used for mixing the slurry should provide for accurate proportioning of lime and water. The proportion of the lime-water slurry depends upon the required percent of lime, the optimum moisture content of the soil or base material, and the field moisture content in the soil at the time of application. During cool weather, care should be exercised to prevent excessive applications of lime slurry which increase the moisture content of the soil. Drying back to optimum moisture content often is very slow and time-consuming. Overwatering the soil during hot weather presents no serious difficulties as the addition of lime usually increases the capacity of the material for water. Slurry consisting of 1 ton of lime and approximately 400 gallons of water (approximately 37 percent solution) is ordinarily the heaviest concentration which can be pumped and spread efficiently. Typical mixtures contain 1 ton of lime and about 500 gallons of water (approximately 32 percent solution). Two or more passes with the tank truck are usually required to spread the lime.

Lime should not be spread on more area than can be satisfactorily mixed during a working day, and can be mixed with soil or base material within 6 hours after application. These requirements are necessary to prevent loss of lime due to wind and rain, and also to prevent the lime from becoming less effective by exposure to the open air. Even though these requirements are met, the spreading operation should be well coordinated with the mixing and compacting operations. When the application of lime is extended too far in advance of the other operations, moisture and density control procedures are complicated by the gradual change in physical properties of the lime-treated mixture.

Lime should not be applied unless the temperature in the shade is at least 40°F and is expected to remain at least 40°F during the mixing period. In no case should lime be applied on a frozen foundation.

307.07 — Mixing Operations

Generally, the same mixing procedure is used for both the dry method and the slurry method of application, except when the materials contain plastic clays or other materials which will not readily mix with lime. The lime must be mixed with the soil, brought to the proper moisture content, sealed and allowed to age for at least 48 hours before the final mixing. Care should be taken during the initial mixing to distribute lime to the proper depth and width in order that the moist clay lumps will be in contact with the lime during the curing period. It is good practice to establish a work schedule which will result in compaction being started on mixed materials which have all been cured about the same length of time.

Final mixing should produce a reasonably uniform blend of lime soil or base material, and water. Moisture contents slightly above the optimum will usually provide better compaction due to the inevitable loss of moisture by evaporation.

307.08 — Compaction, Finishing, and Curing

Compaction of the lime-treated material should begin soon after mixing has been completed.

Although this may depend on the type of roller being used, ordinarily it is advisable to blade the material into windrows on each side of the road to permit rollers to begin compacting at the bottom of the section. When other methods involving thicker lifts are permitted by the specifications, the Inspector should check to ensure that the specified compaction is being obtained in the bottom of the section as well as in the upper portions. When density control methods are specified, the Inspector should be very careful to select representative samples for making the required laboratory compaction tests.

The chemical reaction which usually occurs when hydrated lime comes into contact with moistened clay results in a change in the physical characteristics of the soil. This physical change makes it advisable to make the laboratory compaction tests on the road samples just prior to the time when compaction will start. The laboratory density to be used in the field to check compaction should be determined in accordance with AASHTO T99.

Finishing the lime-treated base course is very similar to the methods used for other flexible-type base courses. If final finishing of the lime-treated base courses involves rewetting and reworking the material, additional lime (1/2 to 1 percent) should be added to maintain strength in the reworked layer. Care should also be taken to use limed-sprinkling water for curing sparingly, on surfaces which will receive a thin asphalt wearing surface. Lime dust coating on the surface sometimes causes difficulty in obtaining good adherence of the asphaltic material to the compacted base course.

The specifications provide for a 7-day period of moist curing. During this period, sprinkling should be done as necessary to maintain the base in a moist condition and to prevent hair-cracking, when asphaltic curing is not used.

307.09 — Cracking and Fluffing

Cracking and fluffing are two objectionable features that may occur in lime-treated base courses, and any construction procedures that minimize or eliminate these problems are beneficial.

It is generally conceded that if compaction of lime-treated materials occurs prior to the chemical changes which lower the plasticity index and change the soils characteristics from plastic to friable, such a procedure may contribute to excessive hair-cracking. This would indicate that compaction should not start until it has been determined by visual inspection that the soil particles have reached a friable condition.

Fluffing is usually associated with the lack of curing, particularly during hot weather. The surface should be kept moist for 7 days after compaction, but repeated flooding of the surface should be avoided. The loss of stability in the top 1/4″ to 3/4″ of the base course is usually attributed to the removal of lime by repeated flooding or excessive manipulation of the surface.

307.10 — Measurement and Payment

Payment for this work is described in the specifications. Measurement of the lime and water, paid by the ton and M.G. (thousand gallons), respectively, must be supported by weight and delivery slips signed by the Inspector when the material is incorporated into the work. If the lime is delivered by the bag, the Inspector should record in the project record the net weight of the bag and number of bags used each day. All entries for payment must be signed and initialed by the person making the entry.

For processing the lime, the station-to-station limits of the completed work should be recorded to determine the length of the treated subgrade.

307.11 — Records and Reports

Daily entries should be made in the diary book describing location of the work, instructions received from the Project Engineer, instructions given to the Contractor, unusual conditions, and other items of interest, similar to those required for Untreated Aggregate Base and Subbase.

For density control projects, a tabulation should be kept of all the density tests made. This tabulation should include the following information: date of test, location of the test hole, (DA) density, estimated moisture content, actual field density, actual moisture content, and percent of (DA) density obtained. Failing test results should be referenced to subsequent tests and notations made as required to explain the action taken.

Section 308 — Cement-Treated Base Course

308.01 — Cement Road-Mix Stabilized Base

Reference: Guide Specification Section 308

308.02 — Description

Cement treated base is normally constructed in one layer and consists of existing or selected soil or aggregate, uniformly mixed in-place with cement and water finishing, curing, and sealing in accordance with the governing specifications and in reasonably close conformity with lines, grade, thickness, and typical cross section shown on the plans. Depths and widths will be shown on the cross sections.

308.03 — Materials

The cement and water should conform to the Guide Specification Section 700, Materials. Water should be free from substances deleterious to the hardening of soil-cement. Water from sources approved for drinking purposes is normally satisfactory.

Soil may consist of existing in-place materials, approved selected material, or a combination of these materials, proportioned as directed within the gradation required by the governing specification.

308.04 — Preparation of Subgrade

Cement stabilized base may be laid upon excavated areas, embankments, or upon imported selected material. The subgrade should be prepared and shaped in accordance with the governing specification. Soft or spongy areas in the subgrade should be removed and replaced with satisfactory material and properly compacted. All drainage conduit, including underdrain if required, shall be in place.

308.05 — Preparation and Preparing of Soils

Selected soil, if used, should be hauled in and placed to an uncompacted depth which, when mixed with cement and water and compacted, will result in a depth, width, and shape required by the typical section.

Existing embankment material may be used in which case it should be determined that the material immediately under the cement stabilized base course is acceptable and compacted to the required density.

After placement of the soil, if imported, the layer should be compacted to the extent that the equipment necessary to distribute the cement and water may be operated over the course. (If embankment material is used, the design depth of the layer should be loosened in the above-described manner with care being taken to assure that the material immediately under the design depth of the stabilized base course is not disturbed.)

308.06 — Preliminary Tests Required

The preliminary design for cement content of the layer is determined by an analysis of the soils available in the area. When the source of the soil is actually established, samples should be forwarded immediately to the Central Laboratory where they will be tested to determine the actual cement content to be used and the optimum moisture for the soil to be used. This testing requires approximately 30 days to produce the desired information before mixing may be started. Accordingly, soil samples of the material to be used should be submitted to the laboratory as soon as possible to minimize delay to the project at this phase of construction.

308.07 — Method of Procedure

The preconstruction conference should include a discussion of the methods of procedure if specified in the contract. If not specified in the contract, the Contractor shall submit the method of procedure for approval. The method of procedure should include methods for controlling traffic, the source of the soil to be used, if known at that time, the paving pattern, the method of pulverizing the soil, distributing water, mixing, shaping, compacting, and curing the layer.

308.08 — Moisture Control of Soil Before Cement is Placed

When the cement is placed on the soil, the moisture content of the soil should be 2 to 3 percent below its optimum moisture as determined by the laboratory. The soils layer should be loosened to the full depth of the layer and thoroughly mixed so that no segregation takes place in the layer.

308.09 — Placement of the Cement

Generally, cement stabilized base is constructed one lane at a time if traffic has to be maintained through the area and also to facilitate the movement of the Contractor's equipment.

The cement shall be placed on the soil in such quantities that the specified cement content of the layer is obtained when mixed. The amount of cement required per square yard should be computed and its placement checked continuously. One method could be the placing of a piece of canvas on a controlled area and weighing the cement deposited on the canvas or by other methods to assure that the cement content of the layer is in close conformity to the design percentage.

Cement-treated material shall not be mixed or placed while the atmospheric temperature is below xxx° (40°F suggested) or when conditions indicate the temperatures may fall below xxx°F (35°F suggested) within 24 hours. Cement-treated material shall not be placed on frozen subgrade or mixed when the aggregate is frozen.

308.10 — Mixing

A major factor affecting the quality of the stabilized base is the thoroughness of mixing of the layer. The soil, cement, and water should be thoroughly mixed throughout the entire depth of the layer. Mixing of the layer may be accomplished either by the mixed-in-place or a central plant method. The mixed-in-place method will be described first.

After the cement has been placed it shall be mixed with the soil. Mixing shall continue until the cement has been blended with the soil sufficiently to prevent the formation of cement balls when water is added.

Immediately after the soil and cement have been mixed, water shall be added and mixed sufficiently to bring the water content to a maximum of 5 percent above its optimum moisture. Generally, the water is spread at three equal distributions of the total water required. To prevent water from running off of the layer, sufficient equipment should be available to complete the wetting and compaction of the layer within 2 hours of the time water addition is started.

Tests should be continuously taken on the wetted mixture to ensure that the desired moisture is incorporated into the mixture. At least one set of test cylinders shall be cast for each day of mixing. After all mixing water has been applied, mixing shall continue until a uniform and intimate mixture of the soil, cement, and water has been obtained, with a suitable mixture resulting. The color of the mixture will be predominately the color of the aggregate.

308.11 — Compaction

At the start of compaction the percentage of moisture in the mixture, based on oven dry weights, shall not be more than 5 percentage points above the optimum moisture content and shall be less than that quantity which will cause the soil-cement layer to become unstable during compaction and finishing. Prior to the beginning of compaction, the mixture shall be in a loose condition for its full specified depth. The loose material shall be compacted to the specified density requirements within 2 hours of the time water addition is started.

Compaction can be accomplished by sheeps foot, wobbly wheel pneumatic-tired, steel-wheel or pneumatic rollers in the combination required to produce the required density.

During compaction operations, shaping will be necessary to obtain uniform compaction and the required grade and cross section.

308.12 — Finishing

During and after compaction the surface of the layer shall be shaped to required lines, grade, and cross section. All roller imprints and other marks made by equipment shall be removed. The resulting surface shall be compacted to the specified density and broom dragged as required.

The specified moisture content of the surface material must be maintained at its specified optimum moisture content during finishing operations. This is usually accomplished by use of a fog spray. Compaction and finishing shall be done in such a manner as to produce, in no longer than 2 hours, a smooth dense sandpaper-like surface free of compaction planes, cracks, ridges, or loose material.

308.13 — Curing

After the treated base has been completed as specified above, it must be protected from drying for 7 days. This is usually done by surface application of asphaltic material on all exposed surfaces. The asphaltic material should be applied as soon as possible, but no later than 24 hours after completion of finishing operations. The finished layer should be kept continually moist until the curing material is placed. The asphaltic curing material must be maintained and replaced if necessary by the Contractor during the 7-day period so that all soil-cement will be protected from damage to the completed work.

Provisions should be made for protection against freezing for a period of 7 days after the construction has been completed.

The lengths of the sections to be processed should be governed by the Contractor's ability to complete those sections during hours of daylight.

308.14 — Joints

Transverse: At the end of the day's run the transverse joint shall be shaped to a vertical plane. All longitudinal joints shall be shaped approximately to a vertical plane. This work is usually done with a grader the day following the finishing of the section. Tapered joints, either longitudinal or transverse, shall not be permitted.

308.15 — Opening to Traffic

Completed portions of cement stabilized base may be opened immediately to traffic of Contractor's equipment if necessary and to all traffic after the 7-day curing period providing the layer has hardened sufficiently to prevent marring or distortion of the surface.

308.16 — Records and Documentation

All pay items shall be listed and computed as described in the contract, and completely documented and recorded in accordance with the departmental policy. A detailed diary must be kept by the Inspector containing the location of work, equipment, source of material, measurements, and all other details pertinent to the work.

308.17 — Cement Plant Mix Stabilized Base

Reference: Guide Specification Section 308

308.18 — Description

Cement plant mix stabilized base consists of existing or selected soil or aggregate, cement and water uniformly mixed in a central batch plant or continuous plant, hauled and spread in one layer, compacted, finished, and cured to the lines, grades, thickness, and typical cross sections shown on the plans.

The comments contained in preceding headings under Materials, Preparation of Subbase, Method of Procedure, Equipment, Compaction, Finishing, Curing, Joints, Opening to Traffic, and Records and Documentation are applicable to this plant-mixed cement stabilized base section.

308.19 — Mixing

The specifications may permit the utilization of various types of mixing equipment, which is optional with the Contractor. Any of these types are capable of producing a satisfactory product if they are in good condition and are operated within reasonable production limitations.

The most common type of mixer is the pugmill, consisting of revolving blades or paddles on a shaft. There may be one or two shafts in the mixer depending on its capacity. The mixer is charged from the top and discharged through the bottom in the case of batch mixing, and charged at one end and discharged at the other in continuous mixing. The quantity of material in the mixer is controlled by batch weights or volumes in batch mixing and by adjustable vertical gates in continuous mixing.

308.20 — Spreading

The Standard Specifications are rather explicit with respect to spreading the cement-treated base, in that self-propelled mechanical spreaders are required. The material shall be spread within the tolerance of 5 percent of a predetermined rate for the width and thickness being spread.

There are a number of self-propelled mechanical spreaders in use which will satisfy the specifications and it is the Engineer's responsibility to see that the spreader on the job conforms. It is imperative that the settings of the screed be checked and watched as the resulting cross section of the completed pavement will be controlled by this operation.

Section 309 — Portland or Blended Hydraulic Cement Concrete Base Course

This section consists of the placement of concrete pavement as a base course. The materials, proportioning, batching, reinforcement, placement, joints, curing, and sampling and testing are essentially the same as that of concrete pavement. Like concrete pavement, this base course can be constructed with or without reinforcement as specified. Please reference Section 500, Rigid Pavement, except for the following exceptions:

309.01 — Proportioning

Either portland cement or a blended hydraulic cement are allowed as the bonding agent. The Rigid Pavement specification only allows the use of portland cement. A blended hydraulic cement is a portland cement with 1% or greater amount of pozzolanic cement (such as fly ash), inert silicious, natural cement or lime, accelerators, retardant, plasticizers and water proofing agents as specified. The term "hydraulic cement" means that the cement sets and hardens in the presence of water.

309.02 — Admixtures

Chemical admixtures, except for air-entraining admixtures, are not in this specification and should not be placed in the concrete mix.

The fly ash subsection is unchanged and should continue to be followed.

309.03 — Final Strike-Off, Consolidation, and Finishing

Texturing and edging of the surface are not required.

Section 310 — Lean Concrete Base Course

310.01 — Description

This section consists of the placement of a lean concrete mix as a base course. The materials, batching, placement, curing, and sampling and testing are essentially the same as that of concrete pavement.

Rotary drum mixers of either batch or continuous type may be used. In general, these employ the same mixing principle as a concrete paver wherein paddles, studs, or flights are made a part of the revolving drum which affords a mixing action as the drum rotates.

The number and arrangement of the paddles or other mixing devices may not be a specification requirement; however, the specifications may offer control of the mixing action through a prohibition against any "dead areas" in the mixer and by specifying the uniformity of cement distribution in the completed mix.

The separation of aggregates into two sizes for cement-treated base may be mandatory and provides control where uniform quality is desired.

The addition of water in the mixing operation is of importance to the quality of the cement-treated base and the spreading and compacting on the follow mixing. The desirable time for addition of water into the mixing operation can be determined by observation and performance on the specific project, but practice indicates that there is some advantage in delaying the addition of water a couple of seconds to get an initial benefit from a limited amount of dry mixing of cement and aggregate. In the case of continuous mixing, the water can be added as the aggregate and cement progress through the mixer following a preliminary dry mixing. Water added to the mixture by sprays facilitates the mixing operation, improves distribution, offers increased production, and leads to a better spreading and compacting operation. The amount of water added is that necessary for hydration of the cement and subsequent compaction which, in general, is close to the optimum moisture content for compaction.

The addition of cement into the mixture in a uniform manner and in the desired amounts is a cause of concern as it has such an important bearing on the finished product. Proportioning cement by weight in batch mixers is generally the most successful method of assuring that the required amounts are processed.

The feeding of cement on continuous type mixers is generally performed with continuous flight augers or vane feeders. Due to the physical characteristics of cement relative to occupying varying volumes under different conditions, it is difficult to be assured of the accurate delivery of the desired weight percentage. The most successful cement feeders now attempt to deliver cement from a constant head receiver in which the cement is agitated by air or other means to keep it in a uniform condition for delivery to the mixer.

With the advent of the cement titration test, there is now available to the Field Engineer a control test which may be performed in a reasonable time to check the performance of the mixing operation. The specifications include the allowable tolerance for variation in the cement, and operation outside of these tolerances must not be condoned. It is the obligation of the Contractor to control and deliver all ingredients to the mixer within the specified tolerances, and the Engineer shall not take or inherit any responsibilities along this line. The Engineer can assist in the calibration of various features of the plants and should make such checks as are necessary to assure uniform operation and control; however, the responsibility for delivery of a specification product is the Contractor's.

The differences between a lean concrete base course with that of concrete pavement is that there is no reinforcement required and no contraction or expansion joints required. The term "lean" means that less cement is required. Also, a blended hydraulic cement can be used as the bonding agent, and a lower quality of aggregate is used.

In general, the aggregates used in this concrete mix would not meet the Rigid Pavement specifications. For example, the quality of an aggregate could be inadequate for the surface of a concrete pavement because of a lack of skid resistance or an excessive amount of wearing or popping out. The lean concrete mixes are designed to make use of these otherwise rejected aggregates in a concrete base course.

Section 500, Rigid Pavement, of this manual applies to this work with the following exceptions:

A. Proportioning. Either portland cement or a blended hydraulic cement are allowed as the bonding agent. The Rigid Pavement specification only allows the use of portland cement. A blended hydraulic cement is a portland cement with 1% or greater amount of pozzolanic cement (such as fly ash), inert silicious, natural cement or lime, accelerators, retardant, plasticizers and water proofing agents as specified. The term "hydraulic cement" means that the cement sets and hardens in the presence of water.

The proportioning part of Section 500, Rigid Pavement, should not be used. Instead the proportioning shall be determined according to the provisions of Subsection 501.03(A) of the AASHTO Guide Specifications except that the cement content shall be sufficient to obtain a workable mix with a slump of between 1 to 3 inches, an entrained air content of between 4 to 9 percent, and a compressive strength within a range of from 750 psi to 1500 psi at 28 days.

The fine aggregate and course aggregate specifications are not used with this mix, instead the aggregate shall be determined according to the provisions of Subsection 703.05 of the AASHTO Guide Specifications.

B. Admixtures. Chemical admixtures, except for air-entraining admixtures, are not in this specification and should not be placed in the concrete mix.

1. Final Strike-off, Consolidation, and Finishing. Texturing and edging of the surface are not required.
2. Joints and Sealing Joints. Transverse and longitudinal contraction and expansion joints are not required. A construction joint shall be constructed whenever concrete placement operations are interrupted for more than 30 minutes. The joint will be formed by placing a header of sufficient dimensions to create a true vertical face perpendicular to the centerline in alignment (the Engineer will determine whether or not tie bars will be required at this joint).

Section 311 — Lime-Fly Ash Treated Courses

311.01 — Description

The addition of lime and fly ash to various plastic subgrade materials has been shown to improve the overall workability of the material. This improved workability is due to the additives having a cementing reaction with the soil. This reaction reduces the plasticity of the soil.

The reaction of these additives to the soil increases the strength and durability of the material. The beneficial effects are generally attributed to complex chemical reactions that occur within the soil-lime-fly ash-water mixture.

The quantities of lime, fly ash, and water necessary will be determined by tests of samples of the soil to be treated. The rates will be specified on the plans or as directed by the Engineer based on the results of the tests. The depth and width of treatment will be designated on the plans.

311.02 — Materials

Lime used will be dry powder or mixed in a slurry. Fly ash used will be a dry powder. They will usually be delivered to the project in bulk. The Inspector should be sure that a Certificate of Compliance is available stating the material delivered complies with the specification requirements. Soil, soil aggregate, or aggregate shall conform to the requirements provided in the contract. The Inspector should be aware of the highly caustic and corrosive nature of quick lime, and should assure that appropriate preventive and protective measures are exercised by those working with this material.

311.03 — Sampling and Testing

Sampling and testing of the soils to determine the percentage of lime, fly ash, and water to be used should be done in accordance with AASHTO T86 and T87. Methods used to determine the various soils constants should be done as outlined in AASHTO T89 through T93 inclusive. From this information, plus whatever additional is needed, the laboratory will determine the percent of lime and fly ash to be used.

311.04 — Preparation of Roadbed

The subgrade should be checked by the Inspector prior to placing lime and fly ash to be sure the grade and shape are within reasonable conformity to the cross section required. Soft spots, ruts, and grade deficiencies should be corrected by removing, replacing, or regrading where necessary.

311.05 — Equipment

The Inspector should inspect all equipment which may include pulverizers, rototillers, traveling plant, water wagons, graders, rollers (pneumatic-tired and steel-wheeled) and all other equipment specified or proposed to be used, for conformity to the governing specifications.

311.06 — Application of Lime

Lime is usually delivered to the project in self-unloading truck transports. Each truck shipment must have the weight of lime certified on public scales, or the weight of lime must be determined on a set of scales furnished by the Contractor. It is essential that the weight of lime be known prior to its use.

If the Contractor proposes to mix the lime and water in windrows, the material must, after grading, be excavated to the required depths. If the Contractor elects to mix with rototilling type machinery, such removal will not be necessary. In either case, grading and rolling prior to lime application must be done in order to obtain the specified final grade and depth of treatment.

Unless otherwise specified, the Contractor has the option of placing the lime on the road either by the "Dry" or the "Slurry" method.

Lime must be distributed with a reasonable degree of uniformity. Dry lime must not be spread when the wind will cause an excessive loss, or when blowing lime becomes objectionable. In distributing lime, a uniform velocity of the vehicle must be maintained and care exercised to secure lapping of the longitudinal strips.

Frequent checks on the rate of application must be made. The actual area of spread for each shipment should be compared with the desired area of spread, as determined from the specified rate, and such information made a part of the job records.

When the slurry method of application is used, lime and water are mixed and applied through spray bars from tank trucks equipped with agitating equipment to keep the lime in suspension. The procedure used for mixing the slurry should provide for accurate proportioning of lime and water. The proportion of the lime-water slurry depends upon the required percent of lime, the optimum moisture content of the soil or base material, and the field moisture content in the soil at the time of application. During cool weather, care should be exercised to prevent excessive applications of lime slurry which increase the moisture content of the soil. Drying back to optimum moisture content often is very slow and time consuming. Overwatering the soil during hot weather presents no serious difficulties as the addition of lime usually increases the capacity of the material for water. Slurry consisting of 1 ton of lime and approximately 400 gallons of water (approximately 37 percent solution) is ordinarily the heaviest concentration which can be pumped and spread efficiently. Typical mixtures contain 1 ton of lime and about 500 gallons of water (approximately 32 percent solution). Two or more passes with the tank truck are usually required to spread the lime.

Lime should not be spread on more area than can be satisfactorily mixed during a working day, and can be mixed with soil or base material within 6 hours after application. These requirements are necessary to prevent loss of lime due to wind and rain, and also to prevent the lime from becoming less effective by exposure to the open air. Even though these requirements are met, the spreading operation should be well coordinated with the mixing and compacting operations. When the application of lime is extended too far in advance of the other operations, moisture and density control procedures are complicated by the gradual change in physical properties of the lime-treated mixture.

Lime should not be applied unless the temperature in the shade is at least 40°F and is expected to remain at least 40°F during the mixing period. In no case should lime be applied on a frozen foundation.

311.07 — Mixing of the Lime

Generally, the same mixing procedure is used for both the dry method and the slurry method of application, except when the materials contain plastic clays or other materials which will not readily mix with lime. The lime must be mixed with the soil, brought to the proper moisture content, sealed and allowed to age for at least 3 days but not to exceed 21 days before the fly ash is added. Care should be taken during the initial mixing to distribute lime to the proper depth and width in order that the moist clay lumps will be in contact with the lime during the curing period. It is good practice to establish a work schedule which will result in compacting being started on mixed materials which have all been cured about the same length of time.

311.08 — Placement of the Fly Ash

After the curing period, the fly ash and water shall be applied and mixing shall begin immediately.

The fly ash shall be placed on the soil in such quantities that the specified cement content of the layer is obtained when mixed. The quantity of fly ash required per square yard should be computed and its placement checked continuously. One method could be the placing of a piece of canvas on a controlled area and weighing the cement deposited on the canvas or by other methods to assure that the fly ash content of the layer is in close conformity to the design percentage.

Lime-fly ash treated material shall not be mixed or placed while the atmospheric temperature is below xxx° (40°F suggested) or when conditions indicate the temperatures may fall below xxx°F (35°F suggested) within 24 hours. Lime-fly ash treated material shall not be placed on frozen subgrade or mixed when the aggregate is frozen.

311.09 — Mixing of the Fly Ash

A major factor affecting the quality of the material is the thoroughness of mixing of the layer. The soil, fly ash, lime and water should be thoroughly mixed throughout the entire depth of the layer. Mixing of the layer may be accomplished either by the mixed-in-place or a central plant method. The mixed-in-place method will be described.

After the fly ash has been placed, it shall be mixed with the soil. Mixing shall continue until the fly ash has been blended with the soil sufficiently to prevent the formation of fly ash balls when water is added.

Immediately after the soil and fly ash have been mixed, water shall be added and mixed sufficiently to bring the water content to a maximum of 5 percent above its optimum moisture. Generally, the water is spread at three equal distributions of the total water required. To prevent water from running off of the layer, sufficient equipment should be available to complete the wetting and compaction of the layer within 2 hours of the time water addition is started.

Tests should be continuously taken on the wetted mixture to ensure that the desired moisture is incorporated into the mixture. At least one set of test cylinders shall be cast for each day of mixing. After all mixing water has been applied, mixing shall continue until a uniform and intimate mixture of the soil, fly ash, lime and water has been obtained, with a suitable mixture resulting. The color of the mixture will be predominately the color of the aggregate.

311.10 — Compaction

At the start of compaction the percentage of moisture in the mixture, based on oven dry weights, shall not be more than 5 percentage points above the optimum moisture content and shall be less than that quantity which will cause the soil-fly ash layer to become unstable during compaction and finishing. Prior to the beginning of compaction, the mixture shall be in a loose condition for its full specified depth. The loose material shall be compacted to the specified density requirements within 2 hours of the time water addition is started.

Compaction can be accomplished by sheeps foot, wobbly wheel pneumatic-tired, steel-wheel or pneumatic rollers in the combination required to produce the required density.

During compaction operations, shaping will be necessary to obtain uniform compaction and the required grade and cross section.

311.11 — Finishing

During and after compaction the surface of the layer shall be shaped to required lines, grade, and cross section. All roller imprints and other marks made by equipment shall be removed. The resulting surface shall be compacted to the specified density and broom dragged as required.

The specified moisture content of the surface material must be maintained at its specified optimum moisture content during finishing operations. This is usually accomplished by use of a fog spray. Compaction and finishing shall be done in such a manner as to produce, in no longer than 2 hours, a smooth dense sandpaper-like surface free of compaction planes, cracks, ridges, or loose material.

311.12 — Protection and Curing

Compaction of the lime-fly ash treated material should begin soon after mixing has been completed.

Although this may depend on the type of roller being used, ordinarily it is advisable to blade the material into windrows on each side of the road to permit rollers to begin compacting at the bottom of the section. When other methods involving thicker lifts are permitted by the specifications, the Inspector should check to ensure that the specified compaction is being obtained in the bottom of the section as well as in the upper portions. Project plans will indicate what methods are to be used in securing the required compaction. When density control methods are specified, the Inspector should be very careful to select representative samples for making the required laboratory compaction tests.

The chemical reaction which usually occurs when hydrated lime and fly ash comes into contact with moistened clay results in a change in the physical characteristics of the soil. This physical change makes it advisable to make the laboratory compaction tests on the road samples just prior to the time when compaction will start. The laboratory density to be used in the field to check compaction should be determined in accordance with AASHTO T99.

Finishing of the lime-fly ash treated base course is very similar to the methods used for other flexible type base courses. If final finishing of the lime-fly ash treated base courses involves rewetting and reworking the material, additional lime and fly ash (1/2 to 1 percent) should be added to maintain strength in the reworked layer. Care should also be taken to use limed-sprinkling water for curing sparingly, on surfaces which will receive a thin asphalt wearing surface. Lime dust coating on the surface sometimes causes difficulty in obtaining good adherence of the asphaltic material to the compacted base course.

The specifications provide for the maintenance of the lime-fly ash treated course in a satisfactory condition until covered by a subsequent course. During this period, sprinkling should be done as necessary to maintain the base in a moist condition and to prevent hair-cracking.

311.13 — Cracking and Fluffing

Cracking and fluffing are two objectionable features that may occur in lime-fly ash treated base courses, and any construction procedures that minimize or eliminate these problems are beneficial.

It is generally conceded that if compaction of lime-fly ash treated materials occurs prior to the chemical changes which lower the plasticity index and change the soils characteristics from plastic to friable, such a procedure may contribute to excessive hair-cracking. This would indicate that compaction should not start until it has been determined by visual inspection that the soil particles have reached a friable condition.

Fluffing is usually associated with the lack of curing, particularly during hot whether. The surface should be kept moist for 7 days after compaction, but repeated flooding of the surface should be avoided. The loss of stability in the top 1/4″ to 3/4″ of the base course is usually attributed to the removal of lime by repeated flooding or excessive manipulation of the surface.

311.14 — Measurement and Payment

Payment for this work is described in the specifications. Lime and fly ash will be measured by the ton. If the lime or fly ash is delivered by the bag, the Inspector should record in the project records the net weight of the bag and the number of bags used each day.

Aggregate will be measured by the ton or cubic yard and the weight of moisture will be deducted.

When specified as a pay item, water will be measured to the nearest 1,000 gallons by calibrated tanks, distributors, or accurate water meters.

The Inspector must support all payments by signed weight and delivery slips when the material is incorporated into the work.

Processing will be measured by the square yard, station, or mile as provided in the contract. The station-to-station limits of the completed work should be recorded to determine the length of the treated course.

311.15 — Record and Reports

Daily entries should be made in the diary book describing location of the work, instructions received from the Project Engineer, instructions given to the Contractor, unusual conditions, and other items of interest, similar to those required for Untreated Aggregate Base and Subbase.

For density control projects, a tabulation should be kept of all the density tests made. This tabulation should include the following information: date of test, location of the test hole, (DA) density, estimated moisture content, actual field density, actual moisture content, and percent of (DA) density obtained. Failing test results should be referenced to subsequent tests and notations made as required to explain the action taken.

DIVISION 400 — FLEXIBLE PAVEMENTS

Section 401 — Plant Mix Pavements

401.01 — General

Plant mix asphalt concrete is a mixture of mineral aggregate and asphalt binder heated and mixed in a stationary central mix plant. All types of plant mixes, whether open graded, or fine graded, are types of asphalt concrete with essentially the same procedure in mixing and placing.

Asphalt concrete is composed of one or a combination of aggregates such as crushed stone, gravel, slag, manufactured lightweight aggregate, shell, sand, screenings, and mineral filler theoretically graded to progressively fill the voids and mixed with asphalt cement to obtain the desired properties. In comparison with road-mix operations, the central plant-mix method permits closer control of moisture content, temperature, and mixture composition which results in greater uniformity in the product. There are other additional advantages, namely: (1) the asphalt material and aggregate are heated before being mixed; (2) the heated mixture can be handled in cool weather, thus extending the working season; (3) compared to road-mix operations no drying effort is lost to wet weather; and (4) there is less interference with traffic because the plant mixture sets up faster after being laid and the long windrow of partially processed materials is unnecessary.

401.02 — Materials and Design

Samples of aggregate and asphalt material proposed for use on the project shall be tested and trial mixes to determine the job-mix formula. Aggregates and asphalt material will normally be submitted by the Project Engineer to the Central Laboratory for quality tests. The aggregate or combination of aggregates shall be of the type and size specified, and the gradation, when screened and graded in the percentages proposed, shall be within the limits of the master grading specified for that mixture. Generally the job-mix formula is determined by the Contractor in conformity with the specification gradation requirements. This job-mix formula will be submitted for approval; however, when required by the contract provisions the Department shall perform the trial mixes necessary to establish the job-mix formula. Trial mixes shall be performed with the grading selected and varying percentages of the specified asphalt cement. The results of these trial mixes should indicate the job-mix formula that will produce the density, stability, and other desired properties required for the project. These trial mixes may be performed by project personnel, District Laboratory, or Central Laboratory. The job-mix formula will be approved by the District or Central Offices.

The use of additives in asphalt concrete, is expensive; and experiments in the use of any that are considered are usually conducted in the design stage. Additives such as hydrated lime, Type I Portland Cement, or anti-stripping agents may permit the use of marginal aggregate subject to stripping that otherwise could not be used. Additives such as granulated rubber and sulfur, may be proposed in an attempt to prolong the life of the pavement.

401.03 — Plant Operations

A. General. There are, in general, two types of plant operations: (1) the permanent commercial plant with multiple production operations; and (2) the portable plant erected on or near the project to produce the mixture for that project.

There are, in general, three types of asphalt concrete-mixing plants: (1) the batch-mix plant; (2) the continuous-mix plant; and (3) the dryer-drum mixing plant.

At the commercial plant site or as soon as the plant is set up and ready to operate, the Project Engineer should make a thorough examination of equipment for compliance with specifications. He should become familiar with plant features determining the mechanical condition of each component part and examine each component part for compliance with specification and safety requirements. Any deficiencies noted in mechanical condition or specification and safety requirements should be corrected prior to beginning mixing operations. Minimum plant production capacity should be determined by the plant inspection personnel.

B. Project Laboratory. The laboratory building of the size and type required by the specifications shall be so located that the plant operations are in full view of the Department inspection personnel assigned to the laboratory. The testing equipment shall be of the type and in such condition that the laboratory personnel may accurately perform the job control tests required by the specifications. The laboratory personnel shall have a copy of the job-mix formula, mix design, project proposal, standard specifications, pertinent addendums to the specifications, contract provisions, and sufficient forms to record all test reports, materials received, and mixture produced. Some acceptance tests may be performed at the plant or part of these tests may be performed by the District or Central Laboratories on samples submitted by project personnel. Process control testing may be performed by the Contractor or the Department. See Division 700 of this manual for this option description and conditions.

C. Stockpiling of Aggregates. Prior to stockpiling aggregates, the stockpile site must be cleared and leveled. Stockpiles should be separated to prevent intermingling. This may be accomplished with clearly defined stockpiles, bins or by using adequate bulkheads. Bulkheads should extend to the full depth of the stockpiles and should be strong enough to withstand pressures that will be exerted under operating conditions. Aggregates must be frequently checked during stockpiling operations for contamination, segregation, and gradation requirements. The stockpiled aggregates shall be of a size and gradation that, when blended together in the proper proportions, will achieve the gradation of the job-mix formula.

D. Storage of Asphalt Cement Materials. The asphalt cement storage tank(s) should be of sufficient capacity to maintain continuous operation while allowing for some delay in shipments. The tanks shall be equipped with sufficient heating coils to evenly heat and maintain the asphalt cement at the specified temperature. The coils usually are heated by steam, hot oil, or electricity. Coating on the storage tanks internal surfaces or additives to the asphalt cement shall not be allowed unless specifically permitted by contract provisions. During mixing operations, the asphalt cement should be continuously circulated in the feeder system between the tanks and the plant. All pumps and feeder lines shall be properly jacketed and heated to maintain the asphalt cement temperature required. A thermometer shall be located near the charging valve to check the temperature of the asphalt cement at the point of use. When required, the storage tanks shall be calibrated so the quantity of material remaining in them may be measured at any time. Asphalt cement is usually sampled and conditionally accepted at the source; however, provisions should be made to sample the asphalt from a sample valve located in a circulatory feed line to obtain verification samples.

E. Mixing Plant (Batch and Continuous). In the batch-mix plant the aggregates are proportioned by weight and asphalt cement proportioned by weight or volume based on weight. In the drum and continuous-mix plants the aggregates and asphalt cement are proportioned by volume based on weight. The component parts and operations of the plants are essentially the same from the stockpiles to the gates of the graded aggregate bins.

1. *Cold Aggregate Feeder.* The cold aggregate feeder normally used with a portable plant is equipped with four or more bins, adjustable gates, reciprocating and/or belt feeders, and an endless belt to carry the proportioned aggregate to the dryer elevator. A commercial plant may be equipped with separate bins, adjustable gates, and a tunnel and conveyor system. In either system, the gates must be adjusted so that the aggregates, in the proper quantity and size, are delivered through the plant in order to maintain uniform production in accordance with the job-mix formula without overflowing.
2. *Dryer.* From the cold feeder, the aggregate is elevated to the dryer where it is heated and dried to the required temperature and moisture content. The component parts of the dryer are: (1) a rotating cylinder (dryer) usually from 3 to 10 feet in diameter and from 20 to 40 plus feet long; (2) a burner which is either gas- or oil-fired; and (3) a fan which may be considered part of the dust collector system, but its primary function is to provide the draft air for combustion in the drum. The dryer is equipped with longitudinal cups, or channels, called "lifting flights," which lift the aggregate and drop it in veils through the hot gases and burner flame. The slope of the dryer, its speed of rotation, diameter, length, and number and design of flights control the length of time required for the aggregate to pass through the dryer. The

aggregate passes from the dryer to the hot elevator through a discharge chute near the burner end of the dryer. The sensing element of a thermometric instrument should be located in the discharge chute to record or indicate the temperature of the aggregate as it passes from the dryer and to activate the automatic burner controls.

3. *Dust Collector.* The fan exhausts the draft air from the upper end of the dryer and other components of the plant into the dust collector system. The system may be composed of either a baghouse, cyclone collector, wet washer or "scrubber" or some combination. This draft air contains dust particles, vapor, and gases which must be reduced or controlled to levels defined by environmental regulations prior to emission into the atmosphere. The collected dust may be reintroduced into the flow of aggregate or wasted as required by the specifications.
4. *Screening Unit (Batch or Continuous-Mix Plants).* The heated aggregates are elevated, usually by a bucket elevator, to a screening unit, which separates the aggregate into the required sizes and deposits the various sizes into the graded aggregate bins. The screening unit, on most plants, is of the flat table vibrating type, usually equipped with four decks. The size of the screens on the decks varies with the type of bituminous concrete to be produced. The top deck is covered with a scalping screen which removes all oversize material and discharges this material into a reject chute. The screening unit should be cleaned daily and checked for loose or torn screens. (Project specifications may waive the requirement for plant screens allowing screenless operations.)
5. *Graded Aggregate Bins (Batch or Continuous-Mix Plants).* The bituminous plant shall be equipped with the number of aggregate bins required by the specifications. These bins hold the heated and screened aggregates in various size fractions required for the type of bituminous mixture to be produced. The bin partitions must be tight, free from holes, and of sufficient height to prevent intermingling of aggregates. Each bin should be equipped with an overflow pipe that will discharge any excess aggregate from the bin. Bin shortages or excesses should be corrected by adjusting the cold feeder gates, screen sizes or hot bin pull weights, as appropriate. On batch plants the bottom of each bin is fitted with a discharge gate which may be operated manually or automatically and the gate's closure should be positive enough to ensure that no leakage into the weightbox will occur. On continuous-mix plants the gates must be adjusted and locked in place to provide a continuous and uniform flow of material from each bin to the pug mill mixer. Samples of aggregates from these bins may be secured from "gates" or "windows" in the sides of the bins, or by diverting the flow of aggregates from the bins into sampling containers.
6. *Scales.* On the batch-mix plant, a weigh hopper for the aggregate is located directly under the graded aggregate bins. The weigh hopper is suspended on

the weighing mechanism generally equipped with a springless dial scale on which the weight of aggregate from each bin is marked accumulatively so that the last mark will read the total amount of aggregate in each batch. The sequence of weighing from each bin must be strictly observed. The recommended practice is to weigh the coarse aggregate fraction first.

The asphalt cement is usually weighed into an overflow-type bucket suspended on a weighing mechanism with a springless dial scale.

When the plant is ready to operate, the scales should be cleaned, each part carefully checked, and load tested for the accuracy specified by a commercial scale mechanic. Each scale should be equipped with a quick adjustment to zero to compensate for cumulations of dust and asphalt cement during production operations. Mixing plants should have fully automated controls for proportioning and mixing. The plant should have at hand the required number of test weights for frequent testing of all scales.

The Contractor may provide an approved automatic printer system which will print the weight of each material delivered, provided the system is used in conjunction with an approved automatic batching and mixing control system.

7. *Graded Aggregate Bin Control Gates (Continuous-Mix Plants).* Up to the point of discharge from the graded aggregate bins, the functions of the continuous-mix plant and the batch-mix plant are essentially the same. In continuous-mix plants, the proportioning of the separate sizes of aggregate is accomplished through the adjustable gates on the feeder of the gradation unit which deposits the aggregates onto the elevator to be delivered directly to the pugmill. The asphalt cement is delivered to the pugmill through a calibrated metering pump. The aggregate feeder and the asphalt cement pump are geared to a common power source so that proportions of aggregate and asphalt cement remain constant, regardless of variations in power supply.

   Before production begins, a careful calibration of the flow of aggregates from each feeder gate must be made. The rate of flow of aggregate from each bin is determined by weighing the amount discharged at various openings, and computing the quantity delivered per revolution of the feeder drive shaft. A curve is then plotted for each gate, showing pounds of aggregate per revolution against gate opening in inches.

   The Contractor must furnish a copy of the manufacturer's operating instructions, which will show the operating speed of the feeder and the asphalt cement pump delivery rate for the various sprocket sizes. The sprocket size for the asphalt cement pump must be checked for the delivery rate required, by weighing the quantity of material pumped into a container over a carefully timed interval. A thermometer must be installed in the circulating line just ahead of the pump as the temperature of the asphalt cement must be controlled within a very narrow range to control fluctuations in percentages of asphalt cement in the mixture.

8. *Asphalt Cement Control Unit*. Satisfactory means shall be provided, either by weighing or metering, to obtain the proper quantity of asphalt cement in the mix within the tolerance specified. Means shall also be provided for checking the quantity or rate of flow of asphalt cement as it passes into the mixer.
9. *Pugmill Mixer*. After proportioning, the aggregate and asphalt cement are introduced into the pugmill for mixing. The pugmill mixer should consist of twin shafts equipped with paddles for mixing the ingredients into a homogeneous mass. Main parts are the paddle tips, paddle shanks, spray bar, liners, shafts, discharge gate, and heated jacket. Efficient mixing is dependent on the number and shape of the paddle tips, R.P.M.'s of the mixing shafts, length of mixing time, temperature of the combined materials, quantity of materials in the mixer, and specifically the clearance between the paddle tips and liner plates. The paddle tips and liner plates should be checked for excessive wear and when the clearance exceeds the dimension specified, the paddle tips and liner plates shall be replaced.

   The mixers of batch-mix plants and continuous-mix plants are essentially of the same design, except for the variations in arrangement of the paddle tips. In the batch-mix mixer, the materials are dumped into the center of the mixer and the paddle tips are arranged to give an end-to-center mixing or a run-around (figure eight) mixing pattern. The material is held in the mixer the required mixing time and then discharged through the discharge gate into the transporting vehicles. The mixer shall be equipped with an automatic timing device to automatically regulate the dry-mixing and wet-mixing periods and a batch counter to accurately record the number of batches produced. In a continuous-mix pugmill, the materials are introduced in one end of the mixer and the paddle tips are set to transport the materials to the discharge end as the mixing is accomplished. The degree of mixing varies with the depth or volume of material in the pugmill, which can be controlled by: (1) raising the dam on the discharge end of the mixer to hold the material in the mixing unit for a longer period of time at a depth that will intensify the mixing action; and (2) adjusting or reversing the pitch of the paddles to retard movement of material through the pugmill.
10. *Surge/Storage Bins*. Hot asphalt mixtures may be stored in holding bins especially designed for that purpose. Each holding bin shall be inspected to determine acceptance at specific holding times. Acceptance will be based upon the ability of the holding bin to hold and discharge mixtures within the quality criteria specified in the job-mix formula, and free of segregation. Discharge into the surge/storage bins should not be a direct stream but through a batching device (gob-hopper).
11. *Haul Trucks and Scales*. Haul trucks that are used to transport the mixture to the roadbed should be in good operating condition, capable of hauling the mixture without spillage and capable of dumping into the asphalt paver or

windrow sizer. Prior to beginning hauling operations, each truck will be assigned a number. The inside of each truck bed may be lightly coated with a soap detergent solution, limewater solution, or an approved commercial oil emulsion (soluble oils) in the proportions recommended by the manufacturer. If such coatings are applied, truck bodies shall be raised immediately prior to loading to remove excess coating material. The use of diesel fuel, kerosene or other like products as a truck bed cleaning agent is not to be allowed. When required by the specifications, the truck beds must be properly insulated and equipped with suitable covers.

Truck scales or hanging weight hoppers under holding bins of the size and type specified for the project shall be furnished and installed by the Contractor. A commercial scale mechanic should supervise the installation and perform the initial check on the accuracy of the scales. Periodic checks for accuracy, during construction, shall be observed by the Project Engineer in accordance with specification requirements. Accuracy checks should be recorded in a field book as a part of the permanent project records.

12. *Establishing Mix Proportions.* The Project Engineer shall furnish the Plant Inspector a copy of the job-mix formula after it has been approved by the District or Central Laboratories. During the preliminary operating period and prior to securing any samples, the cold feeder gates must be adjusted to deliver the proper proportion of aggregates to the gradation unit. Sieve analysis of the aggregates from each bin of the gradation unit shall be made in order to calculate the bin weights (or gate openings and pump sprockets for a continuous-mix plant) required to produce a mixture that will conform with the job-mix formula. It is important to draw several dry batches through the plant to give the screening unit a chance to operate at normal operating speed before securing samples used in making these sieve analyses. The hot bin weights determined from these screen analyses will be used to start production of the plant. The procedures for establishing bin weights, sample forms, and possible causes of fluctuations in gradation, density and stability results are generally explained in publications governing the design and production of asphalt concrete mixtures.

13. *Inspection During Mixing Operations.* Prior to the beginning of each day's production, the Plant Inspector should check to see that the various gates, scales, timers, etc., are accurately set before mixing begins. In addition, the screening unit, bins, and overflow vents shall be checked and cleaned. The recording thermometer charts for the previous 24-hour period shall be collected, dated, and filed in the project records.

After mixing begins, and throughout the day, the Inspector must make the required job control tests or submit samples to District or Central Laboratories for testing. If possible, an assistant should perform the routine tests leaving the Plant Inspector free to observe all the plant operations at frequent intervals. The experienced Plant Inspector's presence around the plant will contribute much to the production of a uniform mixture.

During the day, the Inspector shall make periodic checks of: (1) aggregate stockpiles, cold feeder gates and overflow vents for any overflow of the graded aggregate bins; (2) temperature of aggregates, asphalt cement, and mixture in trucks; (3) the allowable tolerance in gradation for each hot bin to assure that it is not exceeded; and that the gradation of the finished mixture does not vary from the specification limits; (4) proper dryer operation, (5) weighing and mixing operations; and observing mixture in trucks for uniformity in appearance.

The Inspector shall observe the work of the scale person on the truck scales at the plant, and shall see that the required tests of the scales are performed. The Inspector must see that the haul tickets are properly made out and issued for each truckload of mixture delivered and shall see that the daily totals are promptly obtained, checked, entered on the daily report, and made a part of the permanent job records.

14. *Records and Reports*. The Plant Inspector's records should reflect a complete summary of materials incorporated and plant operations performed on the project which with proper documentation form an unquestionable basis for pay quantities. These records should include a Plant Inspector's daily diary, tests performed on materials by project and other Department personnel, materials received, and measurements of materials used on the project.

    Test reports should be filed, as required, with the District or Central Offices. These reports should reflect the quality and quantity of materials being incorporated in the work.

15. *Plant Inspector's Check List*. For the convenience of the Project Engineer and Inspector, some of the most important details of inspection on production of hot-mix bituminous concrete are listed below:
    a. Make sure all tests are understood.
    b. See that testing tools and equipment are on hand and in good condition.
    c. Inspect all components of the mixing plant; make sure all deficiencies are corrected before mixing is begun.
    d. Check all scales for accuracy periodically; check scales for correct adjustment to zero daily.
    e. See that stockpiled aggregates are kept separate; see that no intermingling occurs at the cold feeders.
    f. Make frequent checks of temperatures of heated aggregate, check for moisture content.
    g. Watch for evidence (dark smoke from plant exhaust and oily coating of aggregate) of incomplete combustion of burner fuel.
    h. Check frequently the temperature of the asphalt cement.
    i. Establish scale settings for batch weights; observe plant operator frequently to see that correct weights are obtained.
    j. Make daily checks of screens, bins, and overflow vents for proper operation.

k. Check an occasional batch to see that it is mixed the required length of time.
l. Make frequent visual inspections of mix leaving plant for evidence of non-uniformity or incomplete mixing.
m. Check temperature of mix frequently.
n. Inspect truck beds before loading; see that beds are free of excess oil and congealed chunks of mix.
o. Check occasionally with Road Inspectors concerning workability and uniformity of mix delivered to the roadbed.
p. Take samples of mix and complete required tests or submit samples to District or Central Laboratories.
q. Make accurate, complete records of all test results, number of batches mixed, asphalt materials used, and other pertinent data.
r. Develop a mental picture of the appearance of the proper mixture. A load in which the mixture "peaks up" more than usual in the haul truck indicates a lean mix, either from too much fine aggregate or not enough asphalt cement or both. A load in which the mixture flattens out indicates a fat mix, either from too much asphalt cement, too much coarse aggregate, or not enough fine aggregate.

F. Dryer-Drum Mixing Plants. The drum plant involves introduction of the asphalt materials and mixing in the dryer. This process greatly reduces the production costs of asphalt mixtures, and has the capabilities of meeting the stringent requirements of the environmental protection agencies.

Five major components comprise a modern, conventional drum-mix plant. These five are: (1) a multiple bin cold feed system, (2) a charging conveyor for feeding the new aggregate into the drum mixer, (3) the drum mixer, (4) an asphalt concrete surge system, and (5) a dust collection system.

The interior of the drum mixer is divided into two zones. The front half of the drum is the radiant heating zone, where the cold, wet aggregate is heated and dried. In this zone, the aggregate is subjected to radiant heat, convective heat from exposure to the hot exhaust gas stream, and conductive heat from one aggregate particle coming in contact with another particle. Special flights inside the drum move the aggregate away from the burner flame and build a veil of aggregate in front of the flame. Uniform density of the aggregate veil maximizes the heat transfer process and the removal of the moisture from the aggregates.

The rear or lower half of the drum is the convection coating zone. Heat transfer in this part of the drum takes place primarily by convection and conduction. Asphalt cement is introduced into the drum through a pipe entering from the rear of the drum. Moisture released from the aggregate causes the asphalt cement to expand and foam. The aggregate particles tumble through the foaming asphalt cement and are coated in the process. The aggregate veil created by the flights in the upper half of the drum mixer protects the asphalt cement and asphalt-coated particles from exposure to the burner flame, thus preventing premature hardening of the asphalt cement. The remaining flights in the drum allow the asphalt-coated aggregate to continue to be heated until the desired mix discharge temperature is obtained. The moisture content of the mixture upon discharge from the mixer should not exceed _______ (0.5-1.5 percent suggested) by weight.

Upon discharge from the drum mixer, the asphalt concrete mixture is carried up a drag slat conveyor or vertical elevator into a surge and/or storage silo. Because a drum-mix plant manufactures mix on a continuous basis, a silo is required to temporarily store the asphalt concrete until it can be loaded in to a hauling vehicle. The surge/storage silo should be equipped with a batcher or other devices to prevent segregation as the mix is discharged into the silo. Either a wet collector or baghouse can be used for dust collection.

401.04 – Road Operations

A. General. The construction of asphalt concrete pavement begins with the delivery to the roadbed of a workable mixture that has been proportioned and mixed in accordance with the specification. The pavement shall be constructed of the type of mixture, number of courses, and at the depth specified.

Prior to delivery of the mixture, the surface to be paved shall be shaped to the correct grade and cross section, and compacted to the density specified.

If the paving operation sequence is not set forth in the specifications, the Contractor should submit a proposed sequence of paving operations to the Project Engineer for approval and once this sequence has been agreed upon, it will not be changed without prior approval.

B. Traffic Control. Traffic control should be thoroughly discussed at the preconstruction conference. At this time, definite traffic control procedures should be established that provide maximum safety for the workers and the traveling public, with the least interruption of the work. All traffic control devices and procedures used to direct traffic through the construction area should be in accordance with the *Manual on Uniform Traffic Contrcl Devices* (MUTCD) and of the type shown on the plans and approved by the Department.

If traffic is to be carried on an unpaved shoulder during paving operations, adequate measures should be taken to prevent blowing dust from becoming a traffic hazard. On a pavement-widening project, the open trench is a traffic hazard and this edge should be properly signed and delineated at all times.

C. Weather. Weather limitations as to temperature and closed seasons for this type of construction shall be in accordance with the specifications. Plant production operations shall be suspended at the imminent approach of and during wet weather.

D. Duties Before Paving Begins. The Road Inspector should be thoroughly familiar with the plans and specifications for the project and have the equipment necessary to check all phases of the paving operations. The Inspector should check with the Contractor's Superintendent on the paving sequence previously agreed to and see that the construction equipment required by the specifications is on the project, and that the barricades, warning signs, and other traffic control equipment are in place.

Prior to paving operations, the Road Inspector should thoroughly check the surface on which the pavement is to be placed. If the surface is a soil or aggregate, a prime coat, when required, shall be applied and thoroughly cured. On some soil bases a light application of water should be applied for dust and/or moisture control. The surface should be checked for correct grade and cross section, and all work areas, depressions or potholes shall be repaired to give a firm and unyielding paving base. When an existing surface is to be resurfaced, the surface shall be cleaned of dirt and other extraneous matter and all weak areas repaired. If a leveling course is to be applied, the existing surface should be checked and the roughest areas marked to receive a pre-leveling course prior to the leveling course. A tack coat, when required, shall be applied to all surfaces. The tack coat, when required, shall be applied to all surfaces. The tack coat should be applied to the width and length to be paved with considerations given to the operation of traffic and other factors. Tack coat application should be limited to a maximum area which will be covered within the same day's paving operation.

The pavement edges shall be marked by a stringline or paver guide line sufficiently in advance to assure paving continuity. These should be set and nailed to the surface at intervals that will permit the line to be held taut and shall be checked to be sure that the proposed pavement width is obtained. When required, an electronic sensor line shall be set and rigidly supported to the required grade. Mobile stringline equipment may be used for longitudinal control when an electronic sensor line is not required.

E. Inspection of Paving Equipment. The Road Inspector should make a personal inspection of the Contractor's paving equipment, checking the condition and adjustment of the component parts of the paving machine and rollers. By making this inspection prior to beginning paving operations, obvious deficiencies in the condition of the equipment may be discussed and corrected, thus avoiding delays once the work is under way and to assure that the best possible surface finish can be obtained. Listed below are some of the more important details the Inspector should check during inspection of the paving equipment:

1. *Paving Machines*. The Inspector should be familiar with the mechanical features on the type of paver to be used on the project, so that an intelligent appraisal of the condition and adjustment of the machine may be made. Handbooks of operating instructions are available from each manufacturer, in which the various adjustments and operating details are shown. The general features to be checked on paving machines are:
   a. On all paving machines the operating motor should be checked for proper governor operation, and motor operating smoothly, without missing.

b. On track-laying machines, track linkage must be correctly adjusted; tracks and pins checked for excessive wear.
c. On pneumatic-tire machines, all tires shall be inflated to the correct pressure, and the chain drives checked for correct adjustment and excessive wear.
d. Tamper bars shall be checked for correct R.P.M., proper clearance from screed, length of stroke, and excessive wear on the tips.
e. Vibrators on the screed, if provided should be checked for proper operation.
f. The Strike-off plate in front of the screed should be checked for proper height above the vibrating screed.
g. Screed plates shall be checked for excessive wear, proper crown and tilt adjustment, and screed heating burner operating efficiently. Screed extensions shall be in the same true plane and flush with screed bottom. Vibrators on vibrating screed shall be checked for proper operations.
h. Grade or thickness controls, manual or automatic, should be checked for proper operations.

Motor graders shall be in good operating condition, the blade must be reasonably sharp, and the control unit capable of holding the blade to an established line and grade.

2. *Rollers*. Steel-wheel rollers shall be checked to determine that the wheels are capable of rolling in a true plane and that they are free from flat spots or ridges. The steering and driving mechanism must be free of excessive play, or backlash; and the motor and driving transmission free from oil leaks. Each roller will be fitted with a water tank connected to spray bars and mats on each wheel. The wetting mats should be checked for excessive wear and the spray bars checked for proper operation.

Pneumatic-tire rollers shall be equipped with smooth tires of equal size, ply, in good condition, and equally inflated. Tire pressures and loading of the roller may be varied to give the desirable ground contact pressures. All wheels should roll true, without wobble or creep.

Vibratory rollers shall be acceptable for bituminous mixture compaction and shall have the capability to exert compaction effort equivalent to steel-wheel and pneumatic-tire rollers with separate controls for energy and propulsion. Vibratory rollers shall be capable of providing a smooth pavement surface, free of ridges, indentations, or other objectionable features. The use of equipment which results in excessive crushing of the aggregate will not be permitted.

The number, weight, and type of rollers furnished must be sufficient to obtain the required compaction while the mixture is in workable condition. The sequence of rolling operations and the selection of roller types must provide the specified pavement density.

3. *Miscellaneous Tools*. A check should be made to see that the Contractor has available on the project an adequate supply of rakes, lutes, shovels, brooms,

and other small tools. The Contractor shall be properly equipped with portable barricades, cones, or other means of protecting the freshly laid mixture from damage by traffic.

F. Spreading and Finishing. The asphalt mixture is usually spread and finished by a self-propelled paver. In irregular areas, the mixture may be spread and finished by hand.

The principal duty of the Road Inspector is to ensure that a pavement is constructed to the correct grade and cross section set forth in the plans, and with the surface texture density and riding surface proposed in the contract. To achieve these results, the Inspector must continually check the surface to be paved on, the mixture in the trucks, the depth of the mat being placed, surface texture behind the machine, rolling operation, and paved surface with straightedge and a stringline for proper crown and smoothness. Other duties of the Road Inspectors are to collect load tickets and record the location and length of spread of each load.

To begin paving operations with a paver, the screed should be heated to the proper temperature and grade controls set to construct the transverse joint; this must be carefully checked to ensure good riding qualities and that it conforms to the tolerance requirements before the paver is allowed to proceed. Particular care should be exercised in setting the thickness control device to assure the spread and crown desired. When matching the edge of a previously laid section of pavement, the paver screed should overlap the existing edge from 1 to 2 inches and the thickness control should be adjusted to leave the material slightly higher than the previously laid section of pavement. Overlapping this edge will force enough material into this area to be sure that the joint is completely filled and moisture proof. The height of the material above the previously laid edge of pavement shall be adjusted so that when the longitudinal joint is properly compacted, the pavement shall be uniform in cross section within the tolerances specified.

As the trucks arrive with the mixture, the Inspector shall collect the load tickets, check for proper completion, and record the weight and location where the mixture is placed. The Inspector should visually check each truckload of mixture for uniformity and randomly check the temperature of the mixture. A truckload of mix may be rejected for one of the following reasons: (1) too hot; (2) too cold; (3) too much asphalt cement; (4) too little asphalt cement; (5) non-uniform mixing; (6) excess coarse aggregate; (7) excess fine aggregate; and (8) excess moisture. A fast means of communication between the roadway operations and plant operations is essential to placing a workable and uniform mix on the road and keeping load rejection to the minimum. The spread should be checked frequently to ensure the proper amount of mixture is incorporated in the pavement.

If necessary, haul trucks shall be equipped with covers or other suitable means to prevent the entrance of moisture or rapid loss of temperature.

The use of any haul truck whose frame comes in contact with the paving machine or which bears down on the paving machine when dumping the mixture shall not be permitted. The result of either or both of these conditions will be a rough surface.

As the paver proceeds, the grade or thickness control device shall be adjusted to give the spread required by the plans; however, over adjustment or excessive adjustment should not be allowed since continuity of operations is essential to securing a good pavement surface. The speed of the paver should be regulated by the plant's production capacity, the number of trucks, and the speed of the paver must be coordinated to minimize paver stopping and starting. By observing the surface texture behind the machine, and checking the surface with a straightedge, a malfunction in the paver or non-uniformity of mixture may be detected. The Inspector must insist on prompt action to locate and correct any trouble that occurs. Some of the most common difficulties encountered together with possible causes are listed below:

1. *Wavy surface (short choppy waves).* Worn or poorly adjusted tracks or drive chains; truck driver setting brakes too tightly; excessive paving machine speed.
2. *Wave survey (long waves).* Excessive variation in amount of mix carried in auger box ahead of tampers or screed; rolling too early; roller operating too fast; over-controlling screed.
3. *Excessively open surface texture.* Improper adjustment of tamper bar; improper speed of tamper bar; screed plate rough or galled; excessive paving machine speed.
4. *Varying surface texture.* Insufficient mixing; over mixing; overheating mixture; dry mixing period too long; improper placement in the silo; paving hopper improperly operated; segregation of mix in trucks; worn or damaged screed plate.
5. *Bleeding patches on surface.* Asphalt not uniformly mixed; excessive moisture in mix; excess asphalt cement; too much tack or prime coat.
6. *Irregular rough spots in pavement.* Roller standing on fresh surface; abrupt reversing of roller; truck backing into paver; poor workmanship at transverse joints. When the paving machine is equipped with an automatic grade control unit, it is essential that the paving crew and Inspector be thoroughly familiar with its operation and adjustments. When this unit malfunctions, it tends to compound its errors; therefore, the paver must be stopped immediately, the pavement corrected, and the malfunction located and corrected before proceeding with paving operations. Essentially the automatic grade control unit divorces the screed from the upward and downward movement of the paver caused by an irregular surface. The floating arms which attach the screed to the paver are controlled by an automatically adjusting grade control device. This unit has a sensor element which travels on a rigidly set or traveling guide line and provides proper control of the screed for grade control. When required, automatic slope control sensors may be provided.

When the pavement is constructed in more than one course, the longitudinal joint should be offset from each preceding course approximately 6 inches with the surface course joints being at the center of the pavement or at the location of lane markings on multi-lane pavements. When an adjacent lane is being placed the screed should overlap the previously laid lane from 1 to 2 inches to ensure enough material to completely fill the joint.

G. Compaction. Rollers of the type and number set forth in the specifications shall be used in compacting the mixture. After the rollers have been checked for type and operating condition, the rolling operations are started using rolling methods that are usually prescribed in the governing specifications.

Rolling of the longitudinal joint should be done immediately behind the paving operations. The breakdown rolling should follow the paver as closely as possible without shoving or cracking the mat or having the mix pick up on the roller wheels. The second, or intermediate rolling, should follow the breakdown rolling as closely as possible and should be done while the paving mix is still at a temperature that will result in required density. The finish rolling should be completed while the material is still workable enough for the removal of roller marks.

When paving in echelon, approximately 3 inches of the edge which the trailing paving machine is using as a guide should not be rolled until the adjacent lane has been placed and the longitudinal joint is rolled. The trailing paving machine should stay as close as practical to the preceding machine. In no case should an exposed edge be allowed to cool for more than 15 minutes without being rolled. The construction of the transverse and the longitudinal joints in all courses should be carefully controlled so that the joints are not rough.

During rolling, the roller wheels should be kept moist with only enough water to avoid picking up the material. Rollers should move at a slow but uniform speed with drive drum or wheels nearest the paver. Changes in direction should be effected gradually and rollers allowed to roll or slowly brake to a complete stop before reversing. When rollers are parked on the mat they should be parked at a 45-degree angle with the centerline so that subsequent rolling operations will remove any depressions resulting from the parked rollers. It is always best to park rollers off the new mat, or on a portion that has cooled. If rolling causes displacement of the material, the affected areas should be loosened at once with lutes or rakes and restored to original grade with loose materials, before being re-rolled. Rollers may be loaded or unloaded and pneumatic roller tire pressures may be varied as required by the nature and depth of the mixture to be rolled.

H. Road Inspector's Check List. Some of the most important details of inspection on construction of hot-mix asphalt concrete pavements are listed below:

1. Check condition and adjustment of paving machines and rollers.
2. See that traffic control is organized and functioning properly; make sure required signs are in place.
3. Check application of tack coat; do not allow tacking of more surface than will be paved each day; be sure adjoining surfaces such as gutter, curbs, manholes, etc., are properly tacked.

4. Examine pavement base, see that required patching and/or pre-leveling is done; make check on paving depths or spread before paving begins.
5. See that paver guide lines are set.
6. Make sure that cold transverse joints are cut back to the vertical before continuing a lane.
7. Check transverse joint for smoothness and appearance.
8. Watch trucks dumping into paver hopper for adverse effect on paver operation.
9. Check temperature of the mixture at time of delivery to the paver and during compaction of the course.
10. Maintain constant inspection of mat behind paver for signs of roughness or non-uniformity of mixture.
11. See that longitudinal and transverse joints are raked and compacted properly.
12. Make frequent checks of the spread yield and depth.
13. Watch rolling operations; see that best rolling sequence is used to fit conditions; watch for excessive speed of rollers. Check and adjust the amplitude and frequency of vibrating rollers.
14. Keep records of truck loads used each day; check with Plant Inspector concerning daily totals.
15. Make sure the job is in good shape before you leave at the end of the day; see that all lights, barricades, etc., are properly placed; see that all signs not required during non-working hours are removed or covered.

401.05 — Records and Reports

The Road Inspector's records should reflect a complete summary of paving operations on the project. These records should include a Road Inspector's diary, load tickets, weather conditions, and the record of loads received with asphalt cement content and location where the load is placed. The Plant Inspector usually files the reports, when required, with the District and Central Offices; however, the Road Inspector must measure the area paved each day and furnish this information to the Plant Inspector, along with the total weight of material placed and numbers of the load tickets when any loads or partial loads have been rejected or wasted. The records of the quantity of material produced and material placed must agree and be properly documented to form an unquestionable basis for pay quantities.

Section 402—Cold Mix Asphalt Pavement

402.01 — General

Asphalt cold mix is a mixture similar to hot asphalt concrete except that the materials are of such a nature that the mixture may be transported, stockpiled, and laid cold. There are many different combinations of materials used in cold asphalt concrete mixtures; however, all combinations may be classified under two general types: one in which the mineral aggregate is mixed with a liquid asphalt cement (either an emulsion or cutback) bitumen; and another in which asphalt cement, or powdered asphalt, is mixed with the mineral aggregate along with a liquefier, which usually consists of an approved primer and additives such as hydrated lime or water. There are some special kinds of cold asphalt concrete which are patented and hence their use is subject to approval of the patent owner and to the payment of a royalty.

Cold mix asphalt pavement mixture is ordinarily used as a leveling course, surface course, resurfacing course on an existing, but worn surface of another type, and used extensively for patching failures in any pavement surface.

402.02 — Design

As the major portion of mixture is produced by commercial plants set up in a permanent location, the design and plant control is usually performed by the District or Central Office Laboratory personnel. Trial mixes shall be made similar to hot mix trial mixes and the results of these trial mixes should indicate the design mix formula that will produce a mixture with the density and stability specified for the project. The grading of the aggregate and percent asphalt cement shall conform to the requirements of the specifications.

402.03 — Materials

A. Mineral Aggregate. The mineral aggregate shall be of the same nature, graded, and stockpiled as that for hot asphalt concrete, except that when a mineral aggregate has shown poor affinity for asphalt material, its use may be rejected or an approved non-stripping agent must be added to the asphalt material.

B. Asphalt Cement. The liquid asphalt material should be either emulsions or cutbacks and remain fluid at air temperature sufficiently long enough to permit the completion of construction operations. Asphalt cements used with a primer and hydrated lime of water should be of the high penetration, low viscosity type.

402.04 — Plant Operations

The plant operations of asphalt cold mix and hot asphalt concrete are identical except in the drying operation and mixing operations. See Section 401 for details.

The temperature of the mineral aggregate at the mixer must be considerably lower for cold mix. The mixing temperature limits will be set forth in the specifications. These temperature limits must be strictly observed to ensure a mixture that will remain in a workable condition from the time it is mixed until it is incorporated in the pavement. The temperature may be controlled by heating and drying the aggregate and then cooling back to the required temperature; or controlling the heat and rate of flow of the aggregate through the plant so that the aggregate will arrive at the mixer properly dried and at the specified temperature.

The graded aggregates are proportioned to the mixer according to the design mix formula and dry mixed until they are uniformly distributed. For the liquid asphalt cement type, the asphalt material should then be added and mixed with the aggregate the specified wet mixing period. For the asphalt cement with liquefier type, as soon as dry mixing period is completed the primer, hydrated lime or water, and asphalt cement shall be added and mixed in a manner that will produce a workable cold mix conforming to specification requirements.

The most commonly used hot mix cold lay mixture of the asphalt cement and liquefier type consists of an asphalt cement of low viscosity, an approved primer, and water. The percentages of these materials used in the mixture are determined by the design mix formula within the maximum and minimum limits set forth in the specifications. The sequence of introducing these materials into the mixer, and the length of mixing time, will be such that a uniform and workable mixture that conforms to the specification requirements is delivered to the project.

402.05 — Road Operations

The road operations are the same as for hot asphalt concrete except as discussed below. (See Section 401).

The cold mix, after being uniformly windrowed on the paving surface in the amount to give the required spread, must be thoroughly aerated to reduce the hydrocarbon volatiles and/or moisture content to specified maximum or less before spreading operations begin. As this type of mixture ordinarily produces a dense pavement, there is very little loss in the hydrocarbon volatiles and moisture content after compaction; therefore, if these are not removed to the level specified prior to compaction, they will tend to overfill the voids which in turn may cause the pavement to become unstable. This aerating process may be accomplished by any combination of equipment as described in the road-mix asphalt pavement Section 403. As the aerating process is largely controlled by the weather, the weather limitations set forth in the specifications must be strictly observed.

When the mixture is ready for spreading, it shall be uniformly windrowed, spread, finished, and compacted in the same manner as road-mix asphalt concrete. No succeeding course shall be applied until the surface has been checked and approved by the Project Engineer.

402.06 — Records and Reports

Details as outlined in Section 401.

Section 403 — Road Mix Asphalt Pavement

403.01 — General

The type of base course or surface course known as road-mix or mixed-in-place construction consists of a combination of mineral aggregate and asphalt cement with the mixing performed on the road by means of a traveling mixing plant, travel mixers, or other equipment that will satisfactorily mix the materials directly on the roadbed. This type of pavement, in addition to its use as an original base course or surface course in new construction, is frequently used as a retread mat over an existing but worn surface of another type.

The construction shall consist of one or more courses placed on a prepared foundation.

403.02 — Materials

A. Mineral Aggregates. A wide variety of mineral aggregates may be used for road-mix construction; in fact, almost any gradation and type of material which is predominantly granular in character will serve satisfactorily.

The aggregates used in this type of construction are: (1) aggregates that may be already in place as an integral part of the graded roadbed or as a previously constructed base or surface course of gravel, crushed rock, or other suitable material; (2) if the existing aggregates are deficient in quality or quantity, new aggregate from an outside source may be hauled to the roadbed and blended with the existing aggregate; (3) one aggregate hauled to the roadbed from an outside source; or (4) two or more aggregates hauled from outside sources and blended on the prepared roadbed.

B. Asphalt Materials. The asphalt material for this type of construction must necessarily be a liquid that will remain fluid at air temperature sufficiently long to permit completion of the construction operations. The materials ordinarily used are cutback asphalts, and emulsified asphalts. The type and grade of the material to be used on any project is generally determined by the characteristics of the aggregate, type of road mixing equipment to be used, and climatic conditions. Cutback asphalts may have limited use because of EPA air emission controls.

C. Additives. The only additive to be used in road-mix construction is an approved primer that may be required, or used, at the Contractor's option. In lieu of aerating and drying the aggregate, the Contractor may use an approved primer. The primer shall permit suitable coating of the wet aggregate and shall prevent the asphalt coating from stripping in the presence of free moisture.

403.03 — Equipment

There is a wide variety of mixing equipment, capable of producing a satisfactory mixture, available for use in road-mix construction. The Contractor normally proposes to use equipment that is required by his method of operation. This equipment shall comply with the governing specifications and shall be approved by the Project Engineer, both as to type and working condition, prior to the beginning of construction operations.

A. Mixers. The three principal types of travel mixers used in road-mix construction are: (1) the travel plant mixer that receives the aggregate into a hopper from trucks, mixes it with asphalt material and spreads the mixture—all in one pass of the machine; (2) the travel plant mixer that takes the aggregate from a windrow, mixes it with asphalt material and deposits the mixture behind the mixer in a windrow; (3) the rotary pulverizer-type mixer with transverse shafts that mix the asphalt material and aggregate with resolving tines under a hood.

The mixing mechanism of the mixer should be examined daily for excessive wear and broken or defective parts. The pressure pump and meter that delivers the liquid asphalt material from the tank truck to the spray bar on the travel mixers (1) and (2) above should be accurately calibrated to deliver the percentage of asphalt material specified for the mixture.

B. Distributor. The distributor that is used to apply the liquid asphalt material when the mixer is not equipped to do so shall be of the type and be so equipped as to conform to the specifications. The distributor must be kept clean and in good working condition so as to deliver an accurately measured quantity of asphalt cement material at the specified rate and temperature.

C. Motor Graders. The motor grader(s) shall be of the size and type necessary to adequately mix the needed quantities of materials on the roadway and shall be in good working condition. The blade must be reasonably sharp and the control unit should be capable of holding the blade to an established line and grade.

D. Rollers. The rollers shall be of size and type specified for the project. On pneumatic-tired rollers, all tires must be of equal size and equally inflated. Tire pressures and the loading of the roller may be varied to give the ground contact pressures that are desirable for that particular mixture. Steel-wheel rollers should be checked for excessive play in the steering and driving mechanisms, flat spots on wheels, and spray bars operating properly with wetting mats in satisfactory condition.

E. Haul Trucks. Haul trucks that are used to transport aggregates from other sources to the roadbed should be in good operating condition, uniform in capacity, and capable of dumping into a spreader box, windrow sizer, or hopper. When aggregate is measured by volume, all trucks shall be accurately measured and numbered and these measurements and numbers recorded in the project records.

F. Scales. Truck scales of the type and size specified for the project shall be finished and installed by the Contractor. A commercial scale mechanic should supervise the installation and perform the initial check on the accuracy of the scales. Periodic checks for accuracy, during construction, shall be performed by the Project Engineer in accordance with specification requirements.

G. Miscellaneous Equipment. In addition to the equipment listed above, other equipment, such as asphalt storage tanks, asphalt heaters, water trucks, disk harrows, tractors, power brooms, windrow sizers, and other auxiliary equipment, may be required for use on the project.

403.04 — Construction Operations

A. Traffic Control. Traffic control should be thoroughly discussed at the pre-construction conference. At this time, definite traffic control procedures should be established that provide maximum safety for workers and the traveling public, and with the least interruption of the work. All traffic control devices and procedures used to direct traffic through the construction area should be in accordance with the *Manual on Uniform Traffic Control Devices* (MUTCD) and of the type shown on the plans and approved by the Department.

If traffic is to be carried on an unpaved shoulder during paving operations, adequate measures should be taken to prevent blowing dust from becoming a traffic hazard. On a pavement-widening project, the open trench is an unusually dangerous traffic hazard and this edge should be properly signed and delineated at all times.

B. Weather. Weather limitations as to temperature and closed seasons for this type of construction shall be in accordance with the specifications. The weather should be hot and dry to facilitate the evaporation of moisture and volatiles from the mixture. Extra manipulations of the mixture will be required to remove the moisture and volatiles in cool and humid conditions. Work shall be suspended during wet weather.

C. Preparation of Base and Aggregate. When aggregate in the existing road surface is to be used in the asphalt mixture, the surface should first be scarified lightly and bladed to the correct grade and cross section. If no new aggregate is to be added, the reshaped surface should then be scarified to the depth necessary to provide the required amount of material. The loosened material should then be bladed aside and the understratum shaped to the profile and cross section of the proposed finished surface. Next the understratum shall be scarified, bladed, watered, and rolled to the density specified or until a satisfactory foundation for the mat is prepared. After curing, a prime coat should be applied, as required by the contract provisions.

When new aggregate is to be blended with material from the existing road surface, the process is the same as in the previous case except that following the scarifying of the reshaped roadbed to the required depth the new aggregate should be spread over the loosened existing aggregate in the required amount. The two types of aggregate should then be thoroughly blended by travel mixer or harrowing and blading after which all the aggregate should be bladed to the side and the understratum prepared as a foundation, as in the previous case.

When all new aggregate is to be used in the mixture, it is only necessary to shape and compact the existing road surface to receive the mat and apply a prime coat if specified. When more than one new aggregate is to be used, they should be placed on the prepared surface in the required quantity and blended as in the previous case. Usually, one material is windrowed and spread to the width desired and then one or more new aggregates added on to this aggregate until the required quantity of material is in place and ready for blending operations. If the construction is a retread mat over an existing hard-surface pavement, some patching of the pavement may be necessary and the application of a tack coat will be desirable.

D. Applying Asphalt and Mixing. The aggregate, whether new, blended, or salvaged from the existing road, should be bladed into a windrow along the center or one side of the roadbed. The cross-sectional area should then be measured, the volume computed, and the quantity of asphalt cement required determined. Immediately prior to applying asphalt cement, the aggregate should be checked for moisture. The maximum allowable surface moisture content of the aggregate is normally 2 percent, except where emulsified asphalts are used. If the aggregate is wet, the aggregate should be turned by blades, disk harrows, rotary travel mixer, or otherwise aerated until the moisture content is reduced to 2 percent or as specified. When the aerating process is prolonged and impedes construction progress, the Contractor may elect to use, with the permission of the Project Engineer, an approved additive that will permit suitable coating of the aggregate in the presence of free moisture and shall prevent the asphalt cement coating from stripping.

If the traveling plant method of mixing is to be used, the aggregate should be left in the windrow from which it will be picked up by the machine, fed continuously through the plant, mixed with asphalt cement, and redeposited in a windrow behind the machine ready for aerating, spreading, and compacting. In the mixing machine the desired proportions are obtained automatically through devices which measure both the aggregate and the asphalt cement. The combined materials flow into the mixing chamber where they are processed to uniformity and forced out the rear in a continuous stream by the twin pugmill-type mixer. Another travel plant method of mixing is to receive the aggregate in a hopper from the haul trucks, feed from this hopper continuously through the plant, mix with asphalt cement and deposit behind the machine to the grade and cross section prescribed for that course.

When the mixing is to be performed on the roadbed rather than by a traveling plant, the windrowed aggregate should be spread smoothly and uniformly over part of the roadbed to a width convenient for application of liquid asphalt bitumen. The asphalt material should then be sprayed over the aggregate in two or three separate applications, each approximately one-third the total amount required. Each application should be followed immediately by partial mixing with blades, disk harrows, rotary travel mixer, or other suitable equipment until as little free asphalt material as possible is left. The intervals between the applications depend upon the results of the partial mixing.

After the third application of liquid asphalt and partial mixing is completed, the entire mass should be windrowed and then bladed alternately from one side of the roadbed to the other, or otherwise manipulated by means which will produce equivalent results, until all particles of aggregate are coated with the asphalt cement material and the whole mass has a uniform color. During mixing, attention should be paid to the vertical and horizontal angles of the mold board of the motor grader; this board should be adjusted so that a complete rolling action of the material is obtained when the windrow is manipulated. Also, during the mixing, aerating and spreading operations to follow, care should be taken to avoid cutting into the underlying foundation or contaminating the mixture with earth or other extraneous matter. When specified, the mixing process shall be confined to part of the width or area of the road so as to allow traffic to pass.

Either method of mixing will produce satisfactory results, but the traveling plant method has the following advantages over manipulating the materials on the roadbed: (1) more accurate control of liquid asphalt content is possible; (2) heavier grades of asphalt cement can be used; (3) a more uniform thickness can be obtained; (4) delays caused by inclement weather will be of shorter duration; and (5) the likelihood of partially mixed material getting wet is eliminated.

If, when the mixing process has been substantially completed, the mixture shows an excess, deficiency, or uneven distribution of asphalt cement material, the unsatisfactory condition must be corrected, and then remixed.

If, for some reason, mixing operations are not completed before the end of the day's work or if the operation is interrupted by weather or other conditions, all loose material shall be bladed into a windrow, whether mixing is completed or not, and shall be retained in a windrow until operations are resumed.

E. Laying, Compacting, and Finishing. Before the mixed material is spread on the road for compacting and finishing, it should be checked for moisture content and for the quantity of the volatile portion of the asphalt material remaining in the mixture. If the moisture content exceeds the maximum allowable for aggregate under the contract provisions, the mixture should be aerated sufficiently by manipulation to remove the excess water.

The volatile portion of the asphalt material serves no useful purpose after the mixture has been prepared and placed, and it must be partially removed if the asphalt cement is to develop the natural cementing properties of which it is capable and the mixture stability expected. Dissipation of the volatile material after the mixture has been compacted is a very slow process; consequently, the volatile substance should be eliminated as far as possible before compaction begins. This should be accomplished by continued manipulation until the mixture is just sufficiently workable to permit satisfactory placing.

When the mixture is ready for placing, the windrowed mixture should be moved to one side of the centerline. Approximately one-half of the windrowed material then is split from this windrow and the spreading process begins with the blade carrying this windrow across the centerline to the edge and back to the remaining half of windrowed material. The remaining material is then bladed to the other edge and back. Sufficient material should be left to continue across the centerline to the first edge and then back across the centerline to the second edge completing the laying operation. A pneumatic roller, of the size and type specified, should begin rolling right behind the motor grader on the first spreading pass and should continue to roll until the blading has been completed. This will compact the mixture from the bottom up, mostly eliminating grader tire marks from the surface. The finish rolling should be completed by a steel-wheel roller of the size and type specified. Any loose material that will not compact should be wasted over the side. As a part of the final finishing, the edges should be trimmed to neat lines and the surface straightedged in accordance with the governing specifications.

After one course has been compacted and cured, other courses, as required by the plans, may be placed on it. This operation should be repeated as many times as necessary to bring the road to the cross section and grade shown on the plans.

No uncompacted mixture should be allowed to remain spread on the roadbed overnight or until resumption of operations following suspensions due to weather or other conditions. To avoid contamination, increase in moisture content, or damage by traffic, such loose material should be bladed into a windrow at the end of each day's work and at the beginning of any interruption and retained in a windrow until operations are resumed.

403.05 — Records and Reports

Project records should reflect a complete summary of materials incorporated and construction operations performed on the project and, with proper documentation, form an unquestionable basis for pay quantities. These records should include project diaries, tests performed on materials by project personnel, by District and Central Office personnel, materials received, measurements of materials used on the project, and final plans.

Tests reports and progress reports should be filed, as required, with the District and Central Offices. These reports should reflect the quality and quantity of materials being incorporated in the work.

Section 404 — Tack Coat

404.01 — General

This section will deal with the type of construction contained in Section 404 of the AASHTO *Guide Specifications for Highway Construction*.

It must be assumed that the proper quantities of asphalt material to be used have been established. It is further assumed that during construction, proper adjustments will be made in the quantities due to permissible, slight variations which occur in the materials and which were not considered in the original determination. A uniform application of asphalt material is extremely important in order to achieve the desired results.

404.02 — Descriptions

Tack Coat. A tack coat consists of applying a liquid asphalt material, usually at a specified rate per square yard, upon the existing pavement surface to ensure a thorough bond between the old and new courses. Tack coats are used primarily in connection with the higher types of asphalt concrete pavements.

404.03 — Design and Materials

Prior to use, all materials must be inspected and tested for compliance with the requirements of the specifications.

404.04 — Equipment

See Section 406.04 for the applicable equipment.

404.05 — Road Operations

Surface Cleaning. The existing surface must be cleaned just prior to the applying of the asphalt material. All foreign materials such as paper and mud should be removed and the entire surface should be thoroughly broomed to remove dirt and dust.

Traffic Control. Traffic control should be thoroughly discussed at the preconstruction conference. At this time, definite traffic control procedures should be established that provide maximum safety for the workers and the traveling public, and with the least interruption of the work. All traffic control devices and procedures used to direct traffic through the construction area should be in accordance with the *Manual on Uniform Traffic Control Devices* (MUTCD) and of the type shown on the plans and approved by the Department.

Weather Limitations. An important factor which must be considered when applying a tack coat is the weather. Construction operations should not proceed when the existing surface is wet, when it is foggy, raining, or threatening rain. Temperature limitations should be in accordance with the specifications.

Alignment. The tack coat must be applied to the width and alignment required by the lift of asphalt material to be laid. It is recommended that the tack coat be applied six inches wider than the proposed width of asphalt to be laid. This is necessary to insure that the tack coat exists under the full width of the proposed lift of asphalt material.

Application of Asphalt Material. Just prior to the application of the liquid asphalt material, the quantity of material in the distributor tank should be determined. For this determination, the distributor should be parked off the roadway with the tank in a level position. Also, the asphalt material in the tank should be checked to make certain that it is at the desired application temperature. If it is not, it should be heated to that temperature before it is applied. After application of the asphalt material has been made, the quantity of material remaining in the distributor tank should be determined. For this determination, the distributor should again be parked off the roadway with the tank in a level position.

Close control should be exercised over the operation of the distributor so that the material will be applied uniformly. If the distributor has been properly inspected and adjusted, the material will be applied uniformly in the transverse direction unless one or more of the nozzles become clogged. To obtain the desired uniform rate of distribution in the longitudinal direction, the circulating pump and distributor must be operated at the proper constant speed. The length of spread for each distributor load of material should be determined and marked on the road as an aid to obtaining the desired rate of application.

The tack coat should be applied as near to the time that the asphalt lift is laid as practicable. When using emulsion it will be necessary at all times for the water in the emulsion to evaporate.

404.06 — Records and Reports

The data concerning the construction operations for tack coats should be recorded in sufficient detail in a bound field book so that the necessary progress reports can be prepared and so that the final pay quantities can be determined at some later time. Records must be kept in such detail and such a manner that the final pay quantities can be substantiated by qualified personnel with a minimum of assistance from the project personnel.

Section 405 — Prime Coat

405.01 — General

This section will deal with the type of construction contained in Section 405 of the AASHTO Guide Specifications for Highway Construction.

It must be assumed that the proper quantities of asphalt material to be used have been established. It is further assumed that during construction, proper adjustments will be made in the quantities due to permissible, slight variations which occur in the materials and which were not considered in the original determination.

405.02 — Descriptions

Prime Coat. A prime coat consists of applying a low-viscosity asphalt cement material, usually at a specified rate per square yard, directly upon the surface of a base or foundation course which is to receive some type of asphalt concrete wearing surface. Its purpose is to penetrate the existing surface, to coat and bond any loose mineral particles to the surface, to provide a dust-free surface, and to promote adhesion between the surface and any successive course of treatment.

405.03 — Design and Materials

Prior to use, all materials must be inspected and tested for compliance with the requirements of the specifications.

405.04 — Equipment

See Section 406.04 for the applicable equipment.

405.05 — Road Operations

Repair Defects. On new base courses it seldom will be necessary to repair the surface, since the base course has been constructed to the specified tolerance when finished. Any defects found in the newly constructed base course must be repaired prior to any application of prime coat.

Traffic Control. Traffic control should be thoroughly discussed at the preconstruction conference. At this time, definite traffic control procedures should be established that provide maximum safety for the workers and the traveling public, and with the least interruption of the work. All traffic control devices and procedures used to direct traffic through the construction area should be in accordance with the *Manual on Uniform Traffic Control Devices* (MUTCD) and of the type shown on the plans and approved by the Department.

Weather Limitations. Prime coat should not be applied when the existing surface is wet, when it is foggy, raining, or threatening rain. Temperature limitations should be in accordance with the specifications.

Application of Asphalt Material. Just prior to the application of the liquid asphalt material, the quantity of material in the distributor tank should be determined. For this determination, the distributor should be parked off the roadway with the tank in a level position. Also, the asphalt material in the tank should be checked to make certain that it is at the desired application temperature. If it is not, it should be heated to that temperature before it is placed. After application of the asphalt material has been made, the quantity of material remaining in the distributor tank should be determined. For this determination, the distributor should again be parked off the roadway with the tank in a level position.

Close control should be exercised over the operation of the distributor so that the material will be applied uniformly. If the distributor has been properly inspected and adjusted, the material will be applied uniformly in the transverse direction unless one or more of the nozzles become clogged. To obtain the desired uniform rate of distribution in the longitudinal direction, the circulating pump and distributor must be operated at the proper constant speed. The length of spread for each distributor load of material should be determined and marked on the road as an aid to obtaining the desired rate of application.

The transverse joints on prime coats should be made carefully. This can be done most successfully by starting and stopping each application of asphalt cement material on building paper. Each successive application should overlap the end of the preceding one by 1/2 of an inch in order to avoid a gap in the surface. By using the building paper there will be less chance of creating areas needing to be squeegeed from the surface.

The longitudinal joints for prime coats which are not placed to the full width of the roadway in a single pass should also be carefully controlled. Since it is not practical to use building paper on these joints, it is better to overlap the adjacent passes rather than to have a gap in the surface.

Application of Blotter Material. If the liquid asphalt materials fail to penetrate the surface, and use of the roadway is required, blotter material (natural sand) should be spread in enough quantity to absorb any excess asphalt.

405.06 — Records and Reports

The data concerning the construction operations for prime coats should be recorded in sufficient detail in a bound field book so that the necessary progress reports can be prepared and so that the final pay quantities can be determined at some later time. Records must be kept in such detail and such a manner that the final pay quantities can be substantiated by qualified personnel with a minimum of assistance from the project personnel.

Section 406 — Seal Coat

406.01 — General

This section will deal with the type of construction contained in Section 406 of the AASHTO *Guide Specifications for Highway Construction*.

It must be assumed that the proper quantities of asphalt cement materials and cover aggregates have been established by at least one of several acceptable methods. It is further assumed that during construction, proper adjustments will be made in the quantities due to permissible, slight variations which occur in the materials and which were not considered in the original determination. This section is particularly devoted to promoting the uniform application of asphalt cement materials and the uniform distribution and retention of the cover aggregates.

406.02 — Description

Seal Coat. A seal coat consists of applying an asphalt cement material, at a specified rate per square yard, upon an existing asphalt concrete surface, and immediately placing a single, uniform application of cover aggregate on the freshly applied asphalt cement material. The cover aggregate is then promptly embedded in the asphalt cement material by rolling.

406.03 — Design and Materials

Prior to use, all materials must be inspected and tested for compliance with the requirements of the specifications.

406.04 — Equipment

All equipment proposed for use must be inspected for compliance with the pertinent requirements of the specifications. All material-carrying equipment must be calibrated so that the quantities of material or materials used can be determined. All equipment, by which a material is to be distributed or spread, must be adjusted so that the material will be properly and uniformly placed. The inspection, calibration, and adjustment of the major parts of the various pieces of equipment which may be proposed for use and the determination of important values for the equipment are described as follows.

A. Asphalt Liquid Distributor

Tank. The tank consists of an insulated shell with flues, a thermometer, baffle or surge plates, a manhole, and an overflow pipe. The capacities of distributor tanks vary considerably. All distributors are equipped with a float-type gauge and a measuring stick for determining the quantity of material in the tank. The measuring stick should be marked in increments of not more than 10 gallons. In order to control and check the rate of application and, in some instances, to provide a basis of measurement for payment, the Contractor should be required to furnish calibration data, both signed by a person of recognized authority. The statement should identify the distributor, and give the interior dimensions and a description of the tank. The tank should be inspected to see that it has not been changed from the dimensions and description contained in the certification statement. If the distributor is new or if the notarized statement and calibration data cannot be furnished, it will be necessary to calibrate the tank to relate the depth of material, as determined by the measuring stick, to the number of gallons contained in the tank.

Heating System. The heating system consists of one or two burners and an equal number of heating flues. Each burner emits a flame directly into a flue which transfers heat to the asphalt cement material. The heating system should be checked to make certain that it is capable of maintaining the asphalt cement material at the desired application temperature. When being heated, the asphalt cement material must be circulated. Care should be taken that the safe maximum heat of the material in the tank is not exceeded.

Circulating System. The circulating system consists of a pump and line passing through the distributor tank to the spray bar and to the hand spray. The pump should be checked to make certain that it is capable of circulating the asphalt material through the tank and the spray bar, and developing and maintaining a constant, uniform pressure along the entire length of the spray bar so that an equal quantity of material will be sprayed from each nozzle without atomizing the asphalt material or emitting a distorted fan.

The control for the valve system, by which the discharge of asphalt material from the nozzles is controlled, should be inspected and adjusted, if necessary. There should be *no* slack in the linkage from the control to the valve system so that *all* of the nozzles will be completely opened or completely closed immediately when the control is operated. The pump tachometer or pressure gauge, which registers the pump discharge, should be checked for accuracy.

Measuring Wheel. A measuring wheel consists of a rubber-tired wheel, mounted on a retractable frame and connected to a dial in the cab of the truck by a cable. The measuring wheel should be checked to determine whether it accurately registers all of the data that it is designed to measure. The wheel should be maintained in a clean condition because if material is allowed to build up and remain on the wheel, erroneous results will be obtained.

Spray Bar. To ensure proper working condition of the spray bar, the following inspections and adjustments must be made.

a Nozzles. The nozzles should be removed from the spray bar, cleaned, and examined for size, wear, and damage to the edges of the nozzle opening. Uniform distribution of the asphalt material depends on the nozzles being in good condition and being the proper size. *Usually, the smallest size nozzle available for a distributor will provide the most uniform distribution.* The nozzles should be set so that the slots are at the angle with the spray bar recommended by the manufacturer of the distributor.

b. Spray Bar Height. The height of the spray bar should be set so that the exact number of laps of bituminous material desired will be obtained. The height for a double lap can be determined by closing every other nozzle, operating the distributor at the proper pump speed or pressure, and raising or lowering the spray bar by not more than 1/2 of an inch at a time until it is determined by visual observation that exactly one single lap of material is being applied. When the closed nozzles are opened, a double lap of material will be applied. For a triple lap, close the second and third, fifth and sixth, etc., nozzles and follow the above procedure.

c. Spread. To ensure uniform distribution, the transverse spread and the longitudinal spread should be checked by any of several acceptable methods. The variation should not exceed 10 percent.

B. Aggregate Spreader

Aggregate spreaders are of three general types: tail gate, mechanical, and self-propelled. Of these types, the self-propelled spreader provides the most satisfactory results. It affords close control on traveling speed, can apply the cover aggregate in a continuous and uniform manner, and can stay relatively close to the distributor.

a. Calibration and Adjustments. The aggregate spreader should be calibrated and adjusted in accordance with the manufacturer's recommendations and operating manual.

b. Spread. The transverse spread and the longitudinal spread should be checked to make certain that uniform distribution will be obtained.

c. Operating Speed. The operating speed should always be less than that at which the spreader will lope or undulate.

d. Connecting Hitch. The hitch by which the spreader connects itself to the aggregate trucks should be checked to make certain that it will afford positive connection.

C. Aggregate Trucks

After the trucks proposed for use in hauling the cover aggregate have been inspected for compliance with the specifications and operating condition, each truck should be assigned an equipment number and only that number should be on the truck. No two trucks should be assigned the same number.

a. Calibration. If cover aggregate is to be measured by volume, determine the volume of each truck bed and record the dimensions with the assigned truck identification number. If cover aggregate is to be measured by weight, determine the weight of the empty truck at such frequency as considered necessary. This information *must* be recorded with the truck identification numbers.

b. Connecting Hitch. The hitch by which the truck is connected to the aggregate spreader should be checked to make certain that it will afford positive connection.

D. Scales

Truck Scales. Truck scales of the type and size specified for the project shall be furnished and installed by the Contractor. A commercial scale mechanic should supervise the installation and perform the initial check on the accuracy of the scales. Periodic checks for accuracy, during construction, shall be performed by the Project Engineer or his authorized representative in accordance with the specification requirements.

E. Rollers

Pneumatic Tire Roller. This type of equipment should be checked to determine that it has the desired effective rolling width, the required number of wheels, that it can be loaded to the desired weight, and that the tires are inflated to the pressure necessary to provide the desired ground contact pressure.

Steel-Wheel Roller. If steel-wheel rolling is specified, the rollers should be checked to see that they can be loaded to the desired weight and to determine whether they have the desired rolling width. Each wheel should be examined to make certain that it is free of grooves, that it is not pitted, and that the wheel rims are not worn excessively.

Vibratory Roller. If vibratory rolling is permitted or specified, results of the rolling should be checked to ensure that satisfactory results are obtained. The use of equipment which results in excessive crushing of the aggregate will not be permitted.

F. Power Broom

A power broom shall be used for cleaning the existing surface in preparation for construction and removing excess aggregate from the new surface after the bituminous material has hardened.

G. Secondary Aggregate Distribution Equipment

The use of equipment, such as a drag broom, as a supplement to the aggregate spreader in order to obtain uniform distribution of the aggregate is sometimes necessary to obtain proper keying. The need and use of secondary aggregate distribution equipment may be modified to meet the requirements for local conditions.

H. Single-Pass Surface Treatment Machine

This type of equipment offers one important advantage in that there is no delay between the application of the asphalt cement material and the aggregate. The same calibrations and adjustments necessary for asphalt distributors and aggregate spreaders must be made for this equipment. All calibrations and adjustments should be made in accordance with the manufacturer's recommendations.

406.05 — Road Operations

The construction operations are of the utmost importance in seal coat work. Even the most precise design will be of no value if the construction operations are not properly conducted.

Preparation of Existing Surface. The importance of this step cannot be overemphasized. The riding surface of the seal coat will be no better than the surface on which it is placed.

Repair Defects. On new base courses it seldom will be necessary to repair the surface, since the base course has been constructed to the specified tolerance when finished. For existing asphalt surfaces, however, it will almost always be necessary to make repairs for surface defects. The most common surface defects are raveling, cracks (transverse, longitudinal, alligator, slippage, and shrinkage), broken edges, potholes, corrugations, depressions, bumps, foreign material adhered to the surface, absorbent areas, and flushed or bleeding areas. These repairs should be made well in advance of the construction operations.

Surface Cleaning. The existing surface must be cleaned just prior to the applying of the asphalt cement material. All foreign materials such as paper and mud should be removed and the entire surface should be thoroughly broomed to remove dirt and dust.

Traffic Control. Traffic control should be thoroughly discussed at the preconstruction conference. At this time, definite traffic control procedures should be established that provide maximum safety for the workers and the traveling public, and with the least interruption of the work. All traffic control devices and procedures used to direct traffic through the construction area should be in accordance with the *Manual on Uniform Traffic Control Devices* (MUTCD) and of the type shown on the plans and approved by the Department.

Weather Limitations. Seal coat construction operations should not proceed when the existing surface is wet, when it is foggy, raining, or threatening rain. Temperature limitations should be in accordance with the specifications.

Alignment. All applications of asphalt cement materials must be made to the width and alignment required by the plans. The Contractor should be required to establish a guide, such as a stringline, along one side of the proposed surface, so that if the path of the asphalt cement distributor follows the guide, the asphalt cement material will be applied to the correct width and alignment. The width is not to be assumed as being correct; it must be measured by the Inspector.

Application of Asphalt Cement Material. Just prior to the application of the asphalt cement material, the quantity of material in the distributor tank should be determined. For this determination, the distributor should be parked off the roadway with the tank in a level position. Also, the asphalt cement material in the tank should be checked to make certain that it is at the desired application temperature. If it is not, it should be heated to that temperature before it is placed. After application of the asphalt cement material has been made, the quantity of material remaining in the distributor tank should be determined. For this determination, the distributor should again be parked off the roadway with the tank in a level position.

Close control should be exercised over the operation of the distributor so that the material will be applied uniformly. If the distributor has been properly inspected and adjusted, the material will be applied uniformly in the transverse direction unless one or more of the nozzles become clogged. To obtain the desired uniform rate of distribution in the longitudinal direction, the circulating pump and distributor must be operated at the proper constant speed. The length of spread for each distributor load of material should be determined and marked on the road as an aid to obtaining the desired rate of application.

Liquid asphalt materials cool rapidly; therefore, the distribution of the asphalt material should be coordinated with the spreading of the cover aggregate, if cover aggregate is planned. The time lapse between the distribution of the asphalt material and the application of the cover aggregate should be kept to an absolute minimum to obtain greater wetting action and better seating of the aggregate.

The transverse joints on seal coats should be made carefully so they will not be rough and unsightly. This can be done most successfully by starting and stopping each application of asphalt material and cover aggregate on building paper. Each successive application should overlap the end of the preceding one by 1/2 of an inch in order to avoid a gap in the surface.

The longitudinal joints for seal coats which are not placed to the full width of the roadway in a single pass should also be carefully controlled. Since it is not practical to use building paper on these joints, it is considered better to have a slight build-up due to overlapping the adjacent passes than to have a gap in the surface.

Application of Cover Aggregate. As discussed above under "Application of Asphalt Material," the application of the cover aggregate should be scheduled so that the time lapse before the distributed asphalt material is covered will be the absolute minimum. Care should be exercised in the spreading of the aggregate so that it is spread to a depth of approximately one particle thickness, since more aggregate than this usually will not be stuck and will be wasted. The desired uniform rate of application can be obtained by using an aggregate spreader, if it has been properly adjusted and if it is operated at the proper constant speed, there is no reason for spreading any cover aggregate by hand.

Rolling. The rolling operation should immediately follow the aggregate spreading in order to embed the aggregate while the asphalt material is still soft and tacky. The utmost caution must be exercised in the use of steel-wheel rollers. It is important to avoid excessive rolling. Quite frequently crushing degradation of the aggregate occurs due either to the roller being too heavy or to too much rolling. Usually one coverage with the steel-wheel roller is adequate on all courses of aggregate.

Pneumatic-tired rollers must be operated at a speed slow enough to prevent the tires from displacing or picking up the aggregate. The ground contact pressure may be adjusted by adjusting the amount of ballast on the roller or adjusting the tire pressures, or both. The rolling operation should begin at the outside edge of the surface and progress toward the center. Each pass of the pneumatic roller should overlap the preceding pass by at least one-half of the roller width. Rolling should be discontinued when the bituminous material has set or hardened.

Excess Cover Aggregate. When placing seal coats in half width, the loose aggregate should be removed from along the longitudinal joint before the adjacent lane is surfaced.

Usually, there will be some loose aggregate particles on a new surface after the rolling operation has been completed. This loose aggregate should be broomed off in the cool part of the morning when the asphalt cement material is hard and the bonded aggregate particles will not be disturbed. If traffic is to be returned to the newly covered surface, the surface should be broomed. This is a recommended practice for each half of a roadway that is surfaced in half-widths because the half that is finished first will probably carry traffic while the other half is being surfaced, and damage to automobile finishes and windshields will be minimized. Loose aggregate under traffic may create more loose aggregate.

406.06 — Records and Reports

The data concerning the construction operations for any seal coat should be recorded in sufficient detail in a bound field book so that the necessary progress reports can be prepared and so that the final pay quantities can be determined later. Some type of pay document should be prepared in duplicate, one copy for the Contractor's records and one for the District or Central Offices' records for each distributor load of asphalt material, and for each truckload of cover aggregate. Records must be kept in such detail and such a manner that the final pay quantities can be substantiated by qualified personnel with a minimum of assistance from the project personnel.

A report should be prepared for each day's operation. The information reported should be as follows:

1. The project identification
2. The date
3. The quantities of materials received
4. The quantities of materials used
5. The beginning and ending station numbers for each application
6. The quantity of material in the distributor tank at the beginning and end of each application
7. The rate of application for the materials
8. The temperature of each distributor load of asphalt cement material at the time it was distributed
9. The sources of materials used
10. The amount of rolling done
11. The time the work began and ended
12. The weather conditions
13. The report number
14. The Inspector
15. A summary of any instructions to the Contractor

All reports should be legible, neat, and signed.

Section 407 — Surface Treatment

407.01 — General

This section will deal with the type of construction contained in Section 407 of the AASHTO *Guide Specifications for Highway Construction*.

It must be assumed that the proper quantities of asphalt cement materials and cover aggregates have been established by at least one of several acceptable methods. It is further assumed that during construction, proper adjustments will be made in the quantities due to permissible, slight variations which occur in the materials and which were not considered in the original determination. This section is particularly devoted to promoting the uniform applications of asphalt cement materials and the uniform distribution and retention of the cover aggregates.

407.02 — Descriptions

Single-Course Surface Treatment. A single-course surface treatment is similar to a seal coat (Section 406) except that it is usually applied to prepared base courses and is for the purpose of waterproofing and providing a wearing surface.

Multiple-Course Surface Treatment. A multiple-course surface treatment is very similar to a single-course surface treatment except that the operation is repeated until the desired number of courses is obtained. The maximum size aggregate for each successive course is usually smaller than the preceding course.

Surface Sealers and Rejuvenating Agents. These treatments consist of applying asphalt emulsions, cutbacks, or rejuvenating agents over the surface of new or existing bituminous pavements. Surface sealers are used to seal previous surfaces, to control raveling, and to retard pavement deterioration.

407.03 — Design and Materials

Prior to use, all materials must be inspected and tested for compliance with the requirements of the specifications.

407.04 — Equipment

See Section 406.04 for the applicable equipment.

407.05 — Road Operations

The construction operations for surface treatment are the same as shown in Section 406.05 and should be followed for surface treatment.

407.06 — Records and Reports

The records and reports for surface treatment are the same as shown in Section 406.06 and should be followed for surface treatment.

Section 408 — Slurry Seal (Latex Modified)

408.01 — General

This work shall consist of the application of Slurry Seal (Latex Modified) to an existing surface. The slurry seal shall consist of a mixture of cationic latex modified asphalt emulsion, mineral aggregate, mineral filler (Portland cement), water, and other additives, properly proportioned, mixed, and spread on the existing surface in accordance with the specifications.

408.02 — Materials and Design

A. Aggregate and Mineral Filler. The Inspector is responsible for seeing that proper test reports are on hand prior to starting the work and that the aggregate is in accordance with the project specifications.

B. Latex Modified Asphalt Emulsion. A mixture of water, asphalt, latex, and an emulsifier is referred to as latex modified emulsified asphalt. In this process, the emulsifier is the agent that permits the mixing of two normally immiscible materials.

C. Water. Water is the major factor in controlling the consistency of the finished product. By weight, it normally composes from 4 to 12 percent of the dry aggregate. A set quantity of aggregate and a set quantity of latex modified asphalt emulsion are introduced into the slurry machine mixer. To obtain the proper working consistency of the finished product, the water is increased or decreased.

D. Material Inspection. The inspector should check the materials to be used for the following:

1. Inspect aggregate for foreign materials.
2. Inspect aggregate stockpile areas; the stockpile area should have suitable base and drainage.
3. Inspect cement for lumps.
4. Inspect water if other than potable water supply.
5. Check moisture content of aggregate as required. Twice daily is normal minimum.
6. Inspect asphalt loading area for adequate drainage to avoid asphalt on truck tires.

408.03 — Equipment

Apart from the slurry machine, only a small amount of supporting equipment is required for a complete slurry project. Usually a sweeper front-end loader combination, water truck, asphalt tanker, and various hand tools are all that are necessary. Occasionally, a roller may be required for some special applications.

A. Slurry Machine. Slurry machines are equipped to carry all materials necessary for producing a slurry mixture on the job site and to mix the different ingredients proportionally in a special mixer. The finished slurry is discharged into a spreader box that spreads the slurry on the pavement as the box is pulled behind the slurry machine.

1. *Tanks*. All machines have separate water and emulsion tanks, and the size of these tanks varies with the machine. The tanks can be filled through hatches on top of the equipment and through bottom-loading attachments. The emulsion tank should be inspected regularly for excess build-up of asphalt. The screen in the bottom of the tank should be inspected to insure hardened asphalt does not accumulate or enter the lines. All water lines, filters, and screens should be kept clean. The emulsion tank should be filled from the bottom to minimize foaming and/or disturbing any hardened asphalt materials which may be in the tank.
2. *Bins*. The aggregate is fed from the bin by a belt or auger system through a controlled gate opening. The bin is generally charged with any standard front-end loader.
3. *Slurry Machine Calibration*. Prior to starting, the inspector must inspect the functions of the slurry machine. Important items to be calibrated or checked are:
   a. Calibrate asphalt pump.
   b. Calibrate water meter.
   c. Calibrate mineral filler feeder.
   d. Calibrate aggregate delivery system.
   e. Check aggregate gate setting as required. Twice daily is normal minimum.
   f. Check pugmill for accumulated material and/or excess cleaning solvent.
   g. Calibrate odometer.
   h. Check for and demand uniform delivery of all materials to the mixer. This is extremely important.

B. Spreader Box. The compartmented spreader box is designed to apply the slurry evenly over the pavement. A flexible squeegee is attached to the front, middle, back, and side parts. The squeegees confine the slurry to the box while the back squeegee also acts as a strike-off. A slurry diverter directs the slurry coming from the mixer evenly over the front two compartments. These compartments feed the slurry onto the pavement while the back compartments add or remove slurry as required.

The box is usually extendible from 8 feet to 13 feet at any desired increment. It is raised into a carry position by means of electric, hydraulic, or hand controls depending on the machine. Lateral movement is made by hand or hydraulic controls. The box should be hinged in the center to allow for crown adjustment. The hand or hydraulic controls should be sufficient to obtain a uniform edge line of the slurry.

Located on the rear squeegee are adjustment screws which help control the thickness of the slurry. The thickness of the slurry is chiefly governed by the aggregate size being used, the consistency of the mix, and surface texture of the old pavement.

The inspector should inspect the following items on the slurry box.

a. Check adjustment screws for uniform attack angle on belt.
b. Check burlap for proper width and uniform coverage of surface.
c. Check belts for wear.
d. Replace burlap drag as needed, normally replaced each day.

408.04 — Road Operation

A. General. The machine moves as fast as the slurry can be produced and placed. If the slurry becomes low in the box, the operator signals the driver to slow down until the rate of production can catch up to the spreading rate. This usually takes some adjustment at the start of each new project, but as time progresses, the rate of speed and the rate of production become uniform.

B. Traffic Control. Traffic control should be thoroughly discussed at the pre-construction conference and definite procedures established to be used for the safety and convenience of the traveling public. Barricades and warning signs of the type shown on the plans shall be furnished and erected by the Contractor at locations designated by the Project Engineer.

Due to the relatively fast movement of paving operations, most of the traffic hazards are concentrated around the paving areas. Traffic shall be directed through the construction area with warning signs, flaggers, and pilot trucks or cars in accordance with the *Manual on Uniform Traffic Control Devices,* and in a manner that provides maximum safety for the workers and traffic and the least interruption of the work.

C. Weather Limitations. Weather limitations as to temperature and conditions shall be in accordance with the specifications.

D. Surface Preparation. Prior to application of the slurry, the entire area to be sealed will be cleaned of vegetation, loose materials, mud, etc. This may require the use of power brooms, hand scrapers, and other hand tools. Following the cleaning and immediately prior to the application of the slurry, the surface should be given a light application of water. This application must be uniform and light enough that it does not cause puddles of water to stand on the pavement surface. In the event that small depressions do result in surface puddles, these wet areas should be hand-broomed or squeegeed to remove the excess water.

E. Slurry Placement. It is very important in the actual laying procedure for the Inspector to keep a close watch on the spreader box itself. The homogeneous slurry should roll in one continuous mass. The "roll" will ordinarily be about 12 inches wide. The operator keeps the box in the desired lateral position by hand or hydraulic controls. It is, of course, the responsibility of the driver to drive as straight a line as possible. This is accomplished by a chain that is attached to the front bumper of the truck, and allows the driver to know the approximate location of the edge of the box.

When coming to the end of a pass, the operator should cut the machine off in order to have as little slurry left in the mixer as possible, and none left in the spreader. This, of course, is a judgment factor and depends on a good operator. The operator should cut off the materials for making the slurry in the same order that they were turned on. The operator should be sure that the slurry is evenly distributed over the box to the end of the pass. If slurry is left in the mixer, the operator should cut the mixer off as soon as possible to prevent over-mixing. If slurry is left in the spreader box, the operator should be sure that there are sufficient personnel to remove the excess slurry when he raises the spreader box. At the end of the pass, the operator cuts off the spray bars and raises the spreader into a carry position. It is very important that anytime the operator stops the truck the spray bars are cut off, and just as important that they be restarted when the operation proceeds.

The mixer should be cleaned after each day's operation. The Inspector should make certain the mixer is clean before use. The Inspector should also examine the mixing compartment and blades to assure there is no excessive wear which would cause poor mixing.

F. Curing. There are several factors involved in the curing of slurry. One factor is the thickness of the slurry, the thicker the slurry, the longer the curing period. A characteristic of slurry is to cure from the top down; thus, a crust can form on the surface and give a cured appearance, while the slurry underneath will not be cured. The Inspector should watch for this type of scumming, and not allow any traffic over the surface until the slurry is completely cured.

Another variable factor of slurry curing is climatic conditions, the most important factor being the wind. Wind moving hot air gives a favorable curing condition.

Normal curing periods range anywhere from as low as 30 minutes to as high as 6 hours, but the average is from 2 to 4 hours. In any case, the Inspector must be sure that the slurry is completely cured before allowing traffic to enter the area.

Section 409 — Cold Milling Asphalt Pavement

409.01 — General

Cold milling is the process of removing all and/or portions of an existing asphalt pavement to remove distressed pavement, restore cross-section, improved profile, restore clearances, improve drainage and other reasons. The pavement is removed at locations, depths, widths and in accordance with typical sections indicated in the contract or as directed by the Engineer. This work usually includes removal of the milled material from the highway right-of-way and cleaning the remaining pavement surface suitable for maintaining traffic prior to resurfacing. Unless specified otherwise by the contract, the reclaimed pavement becomes the property of the Contractor.

409.02 — Equipment

Equipment for cold milling includes a self-propelled milling machine capable of removing the existing pavement to the required depth of cut and slope. The milling machine should be equipped with a grade control system which will automatically control the longitudinal profile and cross slope of the milled surface by referencing from one or more skid sensors moving along the pavement surface or from a preset fixed referenced line. The machine should be capable of leaving a uniform surface suitable for maintaining traffic, if necessary, without excessive damage to the underlying pavement structure. The milling machine may be equipped with an integral loading means to remove the material being cut from the roadway and discharge the cuttings into a truck in one operation.

Additional equipment may be necessary to remove the pavement in the area of manholes, water valves, curb and gutter and other obstructions.

The milling equipment should be equipped with a means to effectively limit the amount of dust escaping from the removal operation in accordance with air pollution regulations.

409.03 — Road Operation

The milling operation should be continually checked to determine that the proper depth of milling has been achieved, that the proper profile and cross slope are achieved, and that the surface texture is (a) free from longitudinal ridges, and (b) has a uniform pattern. A change in resulting surface may be achieved by varying the forward speed of the milling machine or the speed of the mandrel.

The milled material may be windrowed behind the miller or directly loaded into hauling units and transported to designated locations in accordance with the project specifications.

The milled pavement surface should be thoroughly cleaned of all loose aggregate particles, dust, and other objectionable material by the use of power brooms, power blowers, power vacuums, or other means. Disposal or wasting of oversize pieces of pavement or loose aggregate material should not be permitted within the right-of-way.

The pavement removal operations should be conducted to effectively minimize the amount of dust being emitted. The operation should be planned and conducted so that it is safe for persons and property adjacent to the work including the traveling public.

At the end of any day's production, a smooth transition should be achieved to the existing pavement.

409.04 — Records and Reports

Project records should reflect a complete summary of the construction operations performed on the project and, with proper documentation, form an unquestionable basis for pay quantities. These records should include project diaries, measurements of materials stockpiled and work completed.

Section 410 — Recycling Asphalt Pavement Material

410.01 — General

This operation usually includes existing asphalt milling the pavement, removal of the pavement from the roadway prior to or after pulverization, processing of material through a central hot mix plant with the addition of virgin material or a modifier, and laydown and compaction. This operation may involve additional heat, depending on the type of materials recycled and the stabilizer used.

Two approaches have been used to size the material prior to recycling in a central hot mix plant. The pavement can be reduced in size in-place and then hauled to the central plant, or the pavement can be removed from the site and sizing can be performed with equipment normally associated with aggregate processing.

Plant sizing can be performed with conventional, fixed, and portable crushing and screening equipment. The pavement is normally ripped and broken to a size suitable to be received by the primary crusher prior to loading onto the haul units. It may be economical to use grid rollers or other types of equipment to produce a properly sized material on the roadway prior to hauling to the central plant.

Equipment to centrally process recycled material can be separated into at least three general categories: (1) direct flame heating, (2) superheated aggregate, and (3) without heat.

Central-plant recycling techniques are different from the other methods of recycling in that the material is removed from the roadway and mixed either cold or hot at a central location. Additional asphalt, recycling agents, cement, lime, aggregate, or other materials may be added at the plant.

410.02 — Materials and Mix Design

The mixture design process for central plant recycling requires the determination of both the type and amount of asphalt material to be used. If asphalt is to be the binder, the Department should determine the need for a modifier recycling agent and/or the quantity of asphalt material. This can be determined only by detailed laboratory testing. Samples of the existing asphalt surface should be obtained and subjected to the Department's standard series of tests.

1. *Recycled Material.* The recycled material shall consist of the existing asphalt pavement (aggregates and asphalts). Normally a maximum of 50 percent can be recycled through a batch plant and 70 percent through a drum mixer.
2. *New Aggregate.* The virgin aggregate shall be added as required to meet job-mix specifications.
3. *Asphalt Modifier.* The asphalt modifier or recycling agent shall be capable of giving desired mix properties.

4. *New Asphalt.* Additional asphalt shall be added to meet job-mix design.
5. *Recycled Mixture.* The recycled mixtures shall be a uniform mixture of recycled material, new aggregate and asphalt (as required), and asphalt modifier (as required) combined to have the properties of a new asphalt pavement composition complying with the specifications.

410.03 — Equipment

1. Asphalt heating and mixing plants designed to produce a uniform mixture within the job-mix tolerances. Many existing hot mix plants can be modified to produce the recycled mixture. These modifications generally involve new cold feed bins, with continuous weighing devices.
2. Asphalt paver that is capable of spreading the mixture to the thickness and width specified, true to the line, grade, and crown shown on the plans.
3. Haul trucks.
4. Asphalt distributor.
5. Rollers.
6. Power broom.
7. Aggregate processing equipment and milling, grinding, or pulverization machines depending on the process used to size material.

The basic description, inspection, and use of this equipment is outlined in Section 401.

410.04 — Plant Operations

Plant operations to centrally hot process the recycled material can be separated into at least two general categories: direct flame heating and superheated aggregate.

1. Direct flame heating is typically performed with a drum mixer wherein all materials are mixed simultaneously in a revolving drum with a flame at one end. Problems with air quality have led to several modifications, such as the addition of heat shields, split feeds, and the like.

   Split feed drum mixers introduce virgin aggregate at the flame end of the drum. At about the midpoint of the drum the reclaimed asphalt pavement material is introduced and is heated by the hot gases as well as by heat transfer from the superheated new aggregate.
2. Superheated aggregate can be used to heat reclaimed asphalt pavement. The superheated aggregate is used to heat the reclaimed or old mixture. Standard batch plants can be used for this approach.

   Plant operations are basically as outlined in Section 401.

410.05 — Road Operations

Procedures and requirements are those basically outlined in Section 401.

410.06 — Records and Reports

Project records should reflect a complete summary of materials incorporated and construction operations performed on the project and, with proper documentation, form an unquestionable basis for pay quantities. These records should include project diaries, tests performed, materials received, measurements of materials used on the project, and final plans.

Test reports and progress reports should be filed, as required, with the District and Central Offices. These reports should reflect the quality and quantity of materials being incorporated in the work.

Section 411 — In Place Cold Recycled Asphalt Pavement

411.01 — General

In-place recycling of old asphaltic concrete is not a new concept.

In-place recycling consists of reworking and the in-place pulverizing of the surface to a depth greater than 1 inch, followed by reshaping and compaction.

In-place recycling techniques are different from the other broad categories of recycling in that all construction operations are performed on-grade or in-place. Additional aggregate, stabilizing binder, and/or a rejuvenating agent may be added to the pulverized old pavement material prior to reshaping and compaction.

411.02 — Materials and Mix Design

For most in-place recycling projects, the materials of the existing pavement are adequate and will meet the job-mix design, when rejuvenators are added. The mixture design process for in-place recycling requires the determination of both the type and amount of rejuvenators to be used. The Central Lab should determine the need for a rejuvenating agent and/or the quantity of asphalt material.

411.03 — Equipment

The Inspector should make a personal inspection of the Contractor's equipment. By making this inspection prior to beginning paving operations, obvious deficiencies in the condition of the equipment may be discovered and corrected, thus avoiding delays once the work is underway and to ascertain that the best possible surface finish can be obtained. Listed below are some of the more important equipment:

1. Milling and pulverizing machines.
2. Distributor or heated tank equipped with metering system.
3. Haul trucks.
4. Laydown machine capable of laying 12′-wide mat with electronic grade control.

5. Rollers—preferably one rubber-tired and one steel-wheel-static or vibratory.

The basic description, inspection, and use of this equipment is outlined in Section 401.

NOTE: The contractor may have equipment that incorporates more than one of the operations and forms a train that would (a) mill the existing pavement to the depth required on the plans, (b) deposit the milled material into a pulverizing/screening unit, (c) deposit the pulverized/sized material into mixing chamber where rejuvenators or asphalt material is added and deposit the properly mixed material in an asphalt paving machine or windrow. The windrowed material may be elevated into a conventional asphalt paving machine, laid and compacted using conventional methods.

411.04 — Road Operations

There are two basic in-place recycling options. If the asphaltic concrete surface is approximately 5 in. or less, specially designed pulverization equipment can be used without preliminary ripping and breaking. For asphaltic concrete surfaces thicker than 5 in., motor graders with scarifiers or dozers with ripper teeth will usually be needed for the initial breakup. Heavy equipment (dozers, rollers, compactors) can be used if additional breakdown is required prior to pulverization.

The types of equipment used must be calibrated to insure proper quantities of liquid asphalt are incorporated with a known amount of graded (pulverized) aggregate. Laying and compaction should be accomplished in the same manner as described in Section 401.

411.05 — Records and Reports

Project records should reflect a complete summary of materials incorporated and construction operations performed on the project and, with proper documentation, form an unquestionable basis for pay quantities. These records should include project diaries, tests performed, materials received, measurements of materials used on the project, and final plans.

Test reports and progress reports should be filed, as required. These reports should reflect the quality and quantity of materials being incorporated in the work.

Section 412 — Surface Recycling

412.01 — General

This operation usually includes reworking the surface of a pavement to a depth of less than 1 inch by heater-planer, heater-scarifier, hot-milling, cold-planing, cold-milling devices, or a combination of these processes. The work involves a continuous, single-pass, multistep process that may also incorporate new materials.

412.02 — Materials and Design

The design of mixtures associated with surface recycling techniques is limited by the nature of the operation.

Heater-scarification operations should be carefully evaluated to ensure that the mixture produced will meet the job-mix design. This can be determined only by detailed laboratory testing. Samples of the existing asphalt surface should be obtained and subjected to the agency's standard series of tests. To meet the specifications, a rejuvenating agent, softer grade of asphalt or additional aggregate may be added to the existing asphalt surface.

Surface milling and grinding operations do not require a mix design unless the millings are to be reused. Standard mix design methods can be used to design mixes using milled material.

412.03 — Equipment

The Inspector should make a personal inspection of the Contractor's equipment. By making this inspection prior to beginning paving operations, obvious deficiencies in the condition of the equipment may be discovered and corrected, thus avoiding delays once the work is underway and to ascertain that the best possible surface finish can be obtained. Some of the more important equipment is listed below:

1. *Heater-Planer.* The heater-planer consists of a mobile heating unit followed by a planing device. The heating and planing devices may be contained in one mobile unit, or may be two pieces of equipment.
2. *Heater Scarifier.* The heater scarifier consists of a mobile heating unit followed by closely spaced scarifier teeth. The heater and scarifier are contained in one unit of equipment.
3. *Milling and Grinding Machines.* Surface milling and grinding equipment should be capable of removing pavement to a depth of at least 5 inches and width from a few inches to 12 feet.
4. *Asphalt Distributor.* (See Section 404.04)
5. *Paving Machines.* (See Section 401.04)
6. *Rollers.* (See Section 401.04)

412.04 — Road Operations

1. *Heater Planing.* Heater planing is used primarily to restore pavement longitudinal grade and transverse cross slope and to correct minor surface irregularities.

   Heater-planer operations usually consist of separate mobile units for heating and planing the pavement followed by rolling. Many times additional new asphalt pavement is applied to the new heater planed surface and rolled simultaneously.

2. *Heater Scarifying*. Operations consist of heating and scarifying the surface, adding additional materials, if required, and rolling.

   The basic heater-scarification-overlay procedure is as follows:

   a. A large, mobile combustion chamber is used to heat the pavement to soften the asphalt surface.
   b. Closely spaced scarifier teeth are then used to plow continuous shallow furrows in the softened materials.
   c. An asphalt overlay is placed.
   d. The overlay is compacted to firmly bond the new overlay to the older pavement structure.

NOTE: Steps a, b and c may be accomplished with one combined machine followed by step d. In some processes steps a and b are combined followed by a compaction of the heater scarified surface followed by a traditional asphalt overlay.

3. *Surface Milling and Grinding*. This operation involves the removal of the surface of a pavement by a hot milling, cold milling, or a cold planing machine. The depth of removal is variable and may be as great as 5 inches in a single pass. The millings or shavings are removed from the construction site.

412.05 — Records and Reports

Project records should reflect a complete summary of materials incorporated in construction operation performed on the project and, with proper documentation, form an unquestionable basis for pay quantities. These records should include project diaries, tests performed, materials received, measurements of materials used on the project, and final plans.

Test reports and progress reports should be filed, as required. These reports should reflect the quality and quantity of materials being incorporated in the work.

Section 413 — Fabric Reinforcement of Asphalt Concrete Pavement Flexible

413.01— General

The application of a fabric reinforcement between asphalt concrete pavement layers serves as a waterproofing and stress relieving membrane. The fabric is normally placed on an existing pavement prior to an asphalt overlay. The fabric reinforcement consists of a single application of asphalt cement which is immediately covered with a layer of fabric. The asphalt mixture is then placed on the fabric prior to opening to traffic.

413.02 — Design and Materials

Prior to use, all materials must be inspected and tested for compliance with the requirements of the specifications.

413.03 — Equipment

a. Asphalt Distributor—See Section 406.04.
b. Fabric Laying Equipment—It is necessary to ensure the fabric laying equipment is capable of handling full or partial rolls of fabric and is capable of laying the fabric in a smooth wrinkle free condition. The fabric smoothness may be supplemented with stiff bristle brooms to help achieve smoothness.

413.04 — Road Conditions

The surface to which the fabric is to be applied will be cleaned of dust, water, oil or other foreign matter. Any surface cracks should be cleaned and filled. Potholes or other surface defects should be repaired prior to applying the fabric.

Prior to laying the fabric an application of liquid asphalt of the type and at the rate required by the specifications should be applied to the surface. Care should be taken in regard to alignment and width of the liquid asphalt.

The fabric should be placed after the liquid asphalt has been applied and before the binder has cooled and lost tackiness. The fabric should be unrolled and placed into the binder with a minimum of wrinkles and as smoothly as possible. The fabric must be broomed to remove air bubbles and maximize fabric contact with the pavement surface. Wrinkles must be cut and laid out flat. If misalignment of the fabric occurs, the fabric should be cut, realigned and jointed as directed by the Engineer. Overlap of fabric at joints should be between 2 and 4 inches. Transverse joints should be shingled in the direction of paving to prevent edge pick-up by the paver. Additional binder must be applied to joints at the rate determined by the Project Engineer by hand spraying or brushing. The reinforcement fabric must be imbedded into the liquid asphalt and bonded to the pavement. Self-propelled pneumatic-tired rollers may be used if determined necessary by the Project Engineer.

After completion of the fabric treatment, the fabric shall be overlaid with a course of asphalt concrete pavement in accordance with the specifications. The asphalt pavement material should be applied in the same manner as described in Section 401. Care must be taken in order not to tear the newly laid fabric. In order to help protect the fabric, hot asphalt material may be hand spread ahead of the laydown machine and delivery truck.

413.05 — Records and Reports

Project records should reflect a complete summary of materials incorporated and construction operations performed on the project and, with proper documentation, form an unquestionable basis for pay quantities. These records should include project diaries, tests performed, materials received, measurements of materials used on the project, and final plans.

Test reports and progress reports should be filed, as required. These reports should reflect the quality and quantity of materials being incorporated in the work.

DIVISION 500 — RIGID PAVEMENT

Section 501 — Portland Cement Concrete Pavement

501.01 — Description

The construction of Portland cement concrete pavement is a highly mechanized operation which requires inspection of a vast quantity of material and a working knowledge of numerous types of equipment. Inspectors assigned to this work should be thoroughly familiar with the specifications, special provisions, construction details, and orders of work. Prior to the start of paving operations, a meeting should be arranged between the Contractor's supervisory personnel and the Engineer with the inspection forces to discuss source of materials, handling of materials, plant site, equipment, methods of operations, and specification requirements. A resume of the meeting should be written by the Project Engineer and copies sent to appropriate persons.

501.02 — Materials

A. Specific Requirements. Refer to Division 700 of the Guide Specifications—Materials for specific requirements of materials to be used in the construction of Portland cement concrete pavement.

B. Handling Materials. The plant should be situated at a location to provide adequate storage facilities for the necessary aggregate stockpiles to be built without overlapping of different materials. Stockpiles should be formed on bases approved by the Project Engineer and should be built by methods which do not cause particle segregation. Stockpiles should be built in layers not exceeding 5 feet in thickness, with each layer being completed before the next one is started. Do not permit conical stockpiles, built by discharging the coarse aggregate at one point, or end-dumping over the sides of the stockpiles. Use of equipment on the stockpile should be held to a minimum to avoid contamination, breakage, and segregation.

Aggregates removed from the stockpiles and placed in the bins should be handled in a manner to prevent segregation, degradation, contamination, and to assure relatively uniform moisture.

Regardless of whether aggregates are tested at the point of production or at the job site, a test report showing numerical results must be available showing that the aggregates are acceptable prior to being incorporated in the work.

Cement and fly ash must be handled and stored in a manner to prevent loss, wetting, or contamination. Cement is usually furnished from pretested bins at the cement plant. If there is no documentation accompanying the shipment of cement indicating that it has been tested, a preliminary test sample must be submitted in accordance with the requirements of AASHTO Designation T127. Different types of cement must not be intermingled or used in place of the type specified.

Potable water is acceptable for use in concrete unless it contains sugar. Water of unknown quality should be tested and found acceptable prior to incorporating in the mix.

The intake end of the pipe or hose used in pumping mixing water from a stream or standing body of water should be covered with wire mesh and so located that no foreign mater will enter. Hauling of mixing water should be done in clean, covered containers. Assurance of using acceptable water is the responsibility of the Slab Inspector for on-the-job mixing.

Admixtures, except air-entraining agent, should not be used without the written permission of the Department. Samples of the proposed admixtures, unless previously approved, must be submitted to the laboratory in advance of use.

C. Admixtures

1. Chemical admixtures should conform to AASHTO Designation M194.

2. Fly ash should conform to ASTM C618. Refer to Subsections 501.02 and 501.03 of the guide specifications for further instructions.

D. Sampling and Testing

1. Concrete Sampling. The importance of obtaining truly representative samples of fresh concrete should be emphasized. Unless the sample is representative, tests results will be misleading. Fresh concrete for testing should be obtained in conformance with AASHTO Designation T141. Time limits allowed for taking samples should be strictly adhered to and samples should be protected from the sun, wind, other sources of rapid evaporation, and from contamination.

2. Tests for Air Entrainment. Air Entrainment tests (pressure method: AASHTO Designation T152) (Gravimetric method: AASHTO Designation T121) should be made from material batched at the start of each day's run. If the air content is not within specification limits, the batch should be discarded and necessary adjustments made in the amount of air-entraining agent. The required number of tests will depend on the uniformity of results. If uniform, satisfactory results are obtained, routine check tests should be made four times a day. More frequent testing is necessary when results show wide variation. The importance of calibrating testing apparatus and determining the aggregate correction factor cannot be overstressed.

If concrete contains slag, expanded clays, shales, or other highly porous aggregates, the volumetric methods, ASTM Designation C173 or AASHTO T196, should be used for entrained air determination.

3. Tests for Consistency (Slump). Slump tests should be performed in accordance with AASHTO Designation T119. Consistency of concrete should be determined each time an air-entrainment test is made and at more frequent intervals as deemed necessary to maintain proper control.

Placement of concrete should not be permitted until tests for air entrainment and consistency have been performed and show that specifications requirements have been met. When routine tests indicate a deviation from the specifications for either one or both, placement should be suspended until adjustments have been made and additional tests show concrete to be within specification limits.

4. Yield. Yield tests should be performed in accordance with AASHTO Designation T121.

5. Cement Content. Cement content should be determined in accordance with AASHTO Designation T121.

6. Density
   a. Plastic Concrete. Testing for the density of plastic concrete should be performed in conformance with AASHTO Designation T121. Consolidation with low slump concrete would generally be by vibration rather than rodding in the container.
   b. Hardened Concrete. Testing for the density of hardened concrete should be performed in conformance with AASHTO Designation T217-81.

7. Aggregate Gradation. Sieve analysis for fine and coarse aggregates should be determined in accordance with AASHTO Designation T27.

8. Moisture Content. Moisture content should be determined in accordance with AASHTO Designation T255.

9. Flexural Strength and Compressive Strength. Making and curing of flexural specimens should be in conformance with AASHTO Designation T23. Testing should be in conformance with AASHTO Designation T97 or other approved method, and at designated intervals.

As an alternate, splitting tensile testing can be performed in accordance with AASHTO Designation T198. Three sets of two companion cylinders are required for each day that pavement is placed. Optional cylinders may be submitted to the designated laboratory for early strength tests.

Making and curring of compressive specimens should be in conformance with AASHTO Designation T23. Testing should be in conformance with AASHTO Designation T22. Compressive specimens are not normally required on concrete pavement work except for unusual conditions or special tests.

See Section 707 of this manual for suggested testing and frequency of tests.

501.03 — Construction Requirements

A. Proportioning

1. Field Proportions. Prior to the start of paving operations, a concrete mix determined by the designated laboratory or as submitted by the Contractor and approved by the Department, will be furnished the Plant Inspector so that proper control can be maintained during batching and mixing operations.

The information furnished should include the following: mix proportions; gallons of water per sack assumed in arriving at mix quantities; specific gravity; rodded weight per cubic foot and absorption of the aggregate; percent of air assumed in arriving at mix quantities; basic weights for a cubic yard of air-entrained concrete.

Two different bases of expressing specific gravity may be used in the calculation of concrete proportions and yield. In the case of bulk-day specific gravity, the aggregate includes no moisture within the particles. The bulk-saturated specific gravity includes the weight of moisture in the aggregate pores. It is important that the specific gravity used is consistent with the moisture condition assumed in expressing the basic batch weights of the aggregates. The bulk-day value should be used if batch weights are based on dry aggregates and bulk-saturated value should be used if batch weights are based on aggregates being in a saturated surface dry condition.

The following sample computations for one cubic yard of concrete are based on aggregates being in a dry condition.

Information furnished by Central Laboratory:

Proportions 1:1.40:3.55
Recommended Water 5.0 gallons per sack of cement
Recommended Air Content 5.5 percent

	Dry Rodded Weight Per Cu. Ft.	**Specific Gravity**	**Absorption (Percent)**	**Dry Wts. Based on 6 Sacks of Cement Per Cu. Yd.**
Cement	94	3.15	—	564
Fine Aggregate	109	2.55	0.9	916
Coarse Aggregate (50 percent fine fraction)	98	2.62	1.5	1,043
Coarse Aggregate (50 percent coarse fraction)	98	2.62	1.5	1,044
Water (30 gal. × 8.33 lbs./gal.)				250

2. Scale Weights. The assumption is made that aggregates have been sampled and tested in the specified manner and following total moistures were determined:

Fine Aggregate	4.5 percent
Coarse Aggregate (Fine Fraction)	2.4 percent
Coarse Aggregate (Coarse Fraction)	0.8 percent

Scale weights of aggregates per cubic yard are computed by the following formula:

Dry weight per cubic yard × (1.000 + Total moisture expressed as a decimal)

Fine Aggregate $916 \times (1.000 + .045) =$
$916 \times 1.045 = 957.22$
Use 957

Coarse Aggregate (Fine Fraction) $1043 \times (1.000 + 0.024) =$
$1043 \times 1.024 = 1068.03$
Use 1068

Coarse Aggregate (Coarse Fraction) $1044 \times (1.000 = .008) =$
$1044 \times 1.008 = 1052.35$
Use 1052

3. Effective Moisture. The gallons of effective moisture in the aggregates may be determined by the following formula:

$$\text{Gallons of Effective Moisture} = \frac{(T - A) \times D}{8.33}$$

Where
T = Percent of total moisture in aggregate expressed as a decimal.
A = Percent of absorption in aggregate expressed as a decimal.
D = Weight in pounds of dry aggregate per cubic yard.
8.33 = Weight per gallon of water.

Fine Aggregate: $$\frac{(.045 - .009) \times 916}{8.33} = \frac{.036 \times 916}{8.33} = 3.96 \text{ (Use 4.0)}$$

Course Aggregate: (Fine Fraction): $$\frac{(.024 - .015) \times 1043}{8.33} = \frac{.009 \times 1043}{8.33} = 1.13 \text{ (Use 1.1)}$$

Course Aggregate: (Coarse Fraction): $\dfrac{(.008-.015)\times 1044}{8.33} = \dfrac{-.007\times 1044}{8.33} = -0.88$ (Use -0.9)

Total gallons of effective moisture per cubic yard:

$$4.0+1.1+(-0.9) = 4.2 \text{ gallons.}$$

The amount of water which may be added at mixer per cubic yard = Recommended gallons per cubic yard − Total gallons of effective moisture in the aggregates.

$$30.0-4.2 = 25.8 \text{ gallons}$$

4. Yield. The theoretical volume of the concrete is determined from the basic weights; the specific gravities of the aggregates and cement, total gallons of water used (effective water plus water added at mixer), and the air content of the concrete. The absolute volume may be computed by the following formula:

Cement and aggregates: $\dfrac{\text{Dry Weight}}{\text{Sp. Gr.}\times\text{Wt/Cu. Ft. of Water}} = \text{cu ft.}$

Water: $\dfrac{\text{Total gallons of water used}\times\text{wt. per gallon}}{\text{wt./cu ft. of water}} = \text{cu. ft.}$

$$\frac{\text{Ab. Vol. of Cement, Aggregates and Water}\times\text{Percent of Air:}}{100-\text{Percent of Air}} = \text{cu. ft.}$$

Assume total gallons of water used to be 29.5 gallons per cubic yard and air content of 5.7 percent. The absolute volume would be as follows:

Cement: $\dfrac{564}{3.15\times 62.4} = 2.87$ cu. ft.

Fine Aggregate: $\dfrac{916}{2.55\times 62.4} = 5.76$ cu. ft.

Coarse Aggregate: (Fine Fraction) $\dfrac{1043}{2.62\times 62.4} = 6.38$ cu. ft.

Coarse Aggregate:
(Coarse Fraction):

$$\frac{1044}{2.62 \times 62.4} = 6.39 \text{ cu. ft.}$$

Water:

$$\frac{29.5 \times 8.33}{62.4} = 3.94 \text{ cu. ft.}$$

$$\text{Subtotal} = 25.34 \text{ cu. ft.}$$

Air:

$$\frac{25.34 \times 5.7}{100 - 5.7} = \frac{144.44}{94.3} = 1.53 \text{ cu. ft.}$$

$$\text{Total:} = 26.87 \text{ cu. ft.}$$

If, after actual operations, it is apparent that yield varies from the original design, due to water requirements or air content, adjustments should be made in the batch quantities. Field adjustments are normally made on the fine aggregate portion of the mix. When this adjustment is of a magnitude where the composition or workability of the mix is materially changed, the designated laboratory should be consulted for a possible redesign.

5. Cement Factor. The cement factor is defined as the cement content in sacks per cubic yard of concrete as determined from the summation of the absolute volumes of all ingredients.

$$\text{Cement Factor} = \frac{(\text{Sacks of cement per batch} \times 27)}{(\text{Yield in Cubic Feet})}$$

$$\text{Cement Factor} = \frac{6 \times 27}{26.87} = 6.03$$

6. Water-Cement Ratio. The water-cement ratio is defined as the total weight of water per weight of cement.

$$\text{Water-Cement Ratio} = \frac{(\text{Total weight (lbs) of water})}{\text{Total Weight (lbs)}}$$

$$\text{Water-Cement Ratio} = \frac{29.5 \times 8.33}{6 \times 94} = 0.43$$

B. Equipment

1. *Batch Plants.* Bins for batching aggregates, or bulk cement, should be tight and constructed in such a manner that there is a free flow of material with no accumulation in the corners, and should be loaded in a way to avoid segregation, contamination, or mixing of different material. Weighing hoppers should be of sufficient size to accommodate the batch being weighed without overflowing or coning against bottom of hopper, and should be constructed to empty completely. Cement should be weighed in a separate hopper from the aggregates. All working parts, such as knife edges, shackles, and weighing arms, should be in good condition, free from avoidable friction and readily accessible for inspection and cleaning, as well as being protected from falling or adhering material. There should be no attachments to scales or weighing hoppers which might restrict the free movement of any part of the weighing mechanism or cause inaccurate weighing during actual operation of the equipment. Cement hoppers should be vented to permit escape of air and should be equipped with a canvas trunk to prevent loss of cement during discharge.

The Contractor is responsible for furnishing all necessary equipment and labor for calibrating scales with weights and the following procedure may be used: balance the scales with no load; apply the test weights and record the scale reading; remove the weights, and place enough material to provide the same scale reading. Leave material on scale and add weights again; record reading; then replace weights with material to get same new scale reading. This operation is repeated until the total weight exceeds that which will be used in batching. Scales that are not accurate within 0.5 percent throughout the range of use should not be used until repaired or adjusted. A quick partial check may be made by adding material to the hopper until the scale registers to within about 300 pounds of the batching weight. Then by adding the test weights, the scales can be checked through the working range.

The sensitivity of the scale should be checked during the calibration test and at least daily during batching operations by applying a small weight and observing if there is movement of the indicator. A weight of five (5) pounds for aggregate scales and of two (2) pounds for cement scales is recommended.

The weighing devices should be balanced and sensitivity checked several times a day, with no load, and checked weekly through the working range using standard weights. Complete calibrating should be repeated after any extended delay and at intervals not to exceed 30 days during continuous operations. Whenever the scales are found to be out of adjustment, the Contractor should be notified to discontinue use of the scales immediately.

When volumetric batching is an acceptable procedure, the Inspector should carefully measure the batching containers provided by the Contractor and determine their exact capacity. The saturated surface-day batch weights of the aggregates should then be converted to an equivalent volumetric measurement. Consideration should be given to the moisture conditions of the aggregate including the bulking effect of the fine aggregate.

All joints and connections should be watertight and valves should close adequately to prevent leakage of the water into the mixer before or after the measuring tank has been discharged.

The water tank or meter is to be calibrated by measuring or weighing sample quantities drawn off for various settings of the device for the full range of water measurements required during mixing operations. The accuracy of measuring the water shall be within a range of error of not over 1 percent. Calibration of water-measuring devices should be repeated when there is an indication of a discrepancy in the quantity of water being dispensed.

Air-entraining agents must be added to the mixing water by means of a mechanical dispenser at the time mixing water is being discharged into the mixing drum. The supply tank should have a gauge which shows the quantity of agent on hand at all times and it is recommended that the level of agent in the tank be maintained as constant as possible. The supply line from the measuring unit to the mixing drum should be of a material that will permit visual observation of the flow of the material. The dispenser should be calibrated through the full working range and should dispense the agent within $\pm$ 3 percent of the desired quantity. Calibration is accomplished by measuring the quantity dispensed for various settings and should be verified at intervals during operation of the mixer to assure that the proper quantity of agent is being dispensed for each batch.

Results of all calibrations, verifications, and sensitivity checks should be recorded and become a part of the permanent record.

2. *Site Mixer.* The water-measuring device and air-entraining agent dispenser should be calibrated and checked in accordance with the procedures noted under the instructions to Concrete Plant Inspector.

Blades in the mixer should be measured for wear and if found to be worn 3/4 inch or more should be removed and replaced with new blades. The majority of wear occurs at the center of the blade with little wear at the ends; therefore, the amount of wear can generally be measured from a stringline or straightedge placed along the length of the blade. The Inspector should have the manufacturer's brochure to show the original arrangement and height of blades. Mixing drum and blades should be free of hardened concrete. Mixing drums must be in a condition to avoid spillage of material or leakage of grout, either on the ground or between compartments in multiple batch mixers.

The loading skip must be adequate in size to hold a full batch without spillage of material and should be clean and smooth so that material is discharged into the mixer drum. Lifting cables should be in good condition to assure proper operation and as a safety factor.

The boom on the site mixer should be of such length that will assure placing the concrete in the required location and should be elevated to the point that the discharge bucket will just clear the forms and/or wire with the gates open. This will keep the free fall of the concrete from the bucket to the subgrade to a minimum. The bucket gates should close tightly to prevent leakage of mortar.

The automatic timing device should be checked and adjusted using a stopwatch while the mixer is loaded. Mixing time shall not be less than 50 seconds nor more than 90 seconds. Four seconds should be added to the specified mixing time if timing starts the instant the skip reaches its maximum raised position. Mixing time ends when the discharge chute opens. Transfer time in multiple drum mixers is included in the mixing time. The device should be locked to assure proper mixing time is maintained. Timing should be checked at least once a day during paving operations.

3. *Truck Mixer.* When truck-mixed concrete is permitted for concrete pavement, all equipment and procedures should be in conformance with AASHTO Designation M157. The following items should be checked:

a. Mixers must be of an approved type in a condition to produce uniform and well-mixed concrete. Mixers should have a legible plate supplied by the manufacturer showing the capacities of the drum and the recommended speed of rotation of the drum or blades. The batch size in relation to the capacities of the equipment must be in conformance with specifications.
b. Water-measuring devices and air-entraining dispenser should be calibrated and checked in accordance with the procedures noted under the instructions to Concrete Plant Inspector.
c. Wash water should be carried in an auxiliary tank. If the wash water is to be used in the succeeding batch, a device should be provided to accurately measure this water.
d. Drums should be checked for excessively worn blades and hardened concrete.
e. Drum or blade speed should be checked to determine that it is within the limits of the specifications.
f. Mixing is controlled either by a specified time, or number of revolutions at a specified revolutions per minute. Regardless of the method used for controlling proper mixing, it shall begin after all ingredients are in the mixer, including water. Close cooperation is required between Plant and Slab Inspectors to assure proper mixing time or number of revolutions is being observed and that concrete is placed within the designated time limit. This is normally accomplished by the issuance of a ticket by a Plant Inspector, a copy of which is sent to the Slab Inspector with the truck driver. Trucks should be equipped with a revolution counter.

4. *Slip-Form Paver.* The slip-form paving equipment must be self-propelled and be capable of placing, spreading, consolidating, screeding, and finishing the freshly placed concrete to the proper pavement elevation and cross section within the specified tolerances. The equipment may be designed to complete all paving operations with one machine or the equipment may consist of a placing machine (mechanical spreader) followed by a separate paver unit. Sliding forms on the paver

unit must be rigidly held together laterally to prevent spreading of the forms. Mechanical floats designed to eliminate small surface irregularities must be utilized as a separate machine in the final finishing operation. The paving equipment must finish the surface in a manner which will minimize hand finishing. Slip-form pavers contain various combinations of all or some of the following components; auger spreader, spud vibrators, oscillating screeds, clary screed, tamping bars, and pan floats. The equipment train should be checked for calibration and satisfactory operation in accordance with the manufacturer's manual before paving is allowed to proceed. Critical features include checking all screeds with a stringline to ensure a true plane or crown, checking the height of the finished pavement elevation, checking the vibrating frequency of the vibrators and screeds, checking the feelers or sensors for sensitivity, and the related stringline for tightness to ensure adequate control of line and grade.

5. *Equipment For Fixed Form Paving*
   a. Mechanical Spreader. The elevation of the bottom of the distributing device and the strike-off, whether screw or plow type, are adjustable. When checking the spreader, the strike-off should be set level with the top of the forms at which time the gauges, visible to the operator, should read zero. The strike-off should then be adjusted for proper thickness and the distributing device adjusted so that some concrete will be carried in front of the strike-off.
   b. Transverse Finishing Machine. The transverse finishing machine has two transverse screeds. Screed-wearing plates that ride on the forms should be checked for signs of wear. Screeds may be checked and adjusted in the following manner: center the screed and lift-off forms; stretch fine wires taut between the forms at the front and back of each screed; place blocks of uniform thickness on top of the wires at each form; lower the screeds. The proper crown is then placed in the screed by measuring between the taut wire and the face of the screed and adjusting the hanger bolts. The front screed should be tilted with the front edge slightly higher. The rear screed should be set flat or with not more than 1/16-inch tilt. When two finishing machines are used, both screeds on the rear machine should have little or no tilt.
   c. Transverse Float Finisher. This machine is carried on a long wheel base frame that rides on the forms and finishes the concrete with transverse oscillating screeds and a stationary float. The front screed normally rides on the forms and may be checked in a similar manner to that described under Transverse Finishing Machine. The second screed and the float do not ride on the forms but are suspended from the frame, therefore, their elevation is much less affected by form irregularities. Both screeds and the float should be adjusted to the proposed cross section. If this machine is included in the paving train, care should be taken when they are in the down position so the

ends of the screeds and the float are set about the elevation of the forms. Small final adjustments are made after start of paving to get the exact cross section and proper surface finish.

d. Vibrators. Must be of sufficient number, size, and spacing to provide required consolidation. The impulses must be checked for compliance with specifications prior to start of work and at least daily during paving operations. Care must be exercised to avoid displacement of reinforcements, joints, and tiebars by vibrators.

e. Subgrade Check Template. This piece of equipment assures that proper thickness of pavement is constructed. The template should be placed on the forms and a taut wire stretched across the forms. The distance to provide the proper cross section is then measured from the wire to the lower limits of the template and the necessary adjustments made. The check template should not be permitted to do any cutting and should be checked at frequent intervals to assure that desired results are obtained.

f. Hand Tools and Auxiliary Equipment. Prior to starting concreting operations, all hand tools and auxiliary finishing equipment should be checked for specification requirements and to assure they are in satisfactory condition.

C. Preparation of Grade. Prior to the start of paving operations, the Inspector should be assured that the subgrade is constructed to the approximate typical section and is of proper density for the full width, including form line, or track path for slip-form paving. It is recommended that the subgrade be constructed slightly higher than fine grade elevation where fixed-form paving methods are used with final adjustments after the forms are set. With slip-form paving equipment, the subgrade and the track path for paver should be at fine grade elevation. The track paths should be graded, checked, and maintained in a smooth compacted condition until the pavement is constructed. Any irregularities will be reflected in the finished surface. When a guide wire is used, it should be supported and tensioned to prevent any measurable sag. Immediately ahead of paving, a final check and adjustment should be made to eliminate any irregularities.

D. Setting Forms. Forms should be checked to ensure that they meet the requirements for dimensions; are clean, oiled, and straight; the face is perpendicular to the base; flanges are not bent; and locking devices are in proper working order. The subgrade for forms is to be cut true to grade, usually from a reference stringline. When forms are set, they should be firmly supported throughout their full length. Pins must be adequate in length to avoid measurable movement under equipment and locked in stake holes. Locking devices must be properly fastened. Width between forms must be correct and proper distance from centerline, and at the correct elevation with a smooth grade line. After forms are set and properly tamped, it is advisable to sight along the top of the forms to detect and adjust irregularities exceeding 1/8 inch from true grade or 1/4 inch from true alignment. If it is necessary to make any adjustment, forms shall be retamped. Approved flexible or curved forms of proper radius shall be used for curves of 100-foot radius or less.

E. Conditioning of Subgrade or Base Course. After forms are set, or track path cut for slip-form paving, the fine grade should be cut with an approved subgrade machine. If payment for the granular base material is on a ton basis, the quantity removed during fine grading operations should be moved ahead or, if not used, deducted from the pay quantity of base material. If the base is low, material should be added and compacted. After final shaping of the subgrade, a steel-wheel roller weighing not less than 5 tons should be used. Immediately ahead of paving operations, the subgrade should be checked with the check template in accordance with specification requirements. In addition, periodic checking should be done by measuring down to the subgrade from a taut string or wire across the forms. These measurements should be recorded. Unless subgrade paper is specified, the subgrade should be uniformly moist, but not muddy, at the time the concrete is placed.

F. Handling, Measuring and Batching Materials. See this subsection in AASHTO Guide Specifications.

G. Mixing Concrete

When concrete is mixed in batch plants, it is the responsibility of the Plant Inspector to assure that it is properly mixed and meets the requirements in regard to slump, air content, uniformity, and desired workability when delivered to the subgrade. The Slab Inspector will have this responsibility for site mixers or truck mixers. Wet and dry batches should be avoided and the slump held to within very narrow limits, normally not exceeding 1/2 inch variation.

The Slab Inspector should see that dry batches are delivered in the proper manner and condition and are dumped in the skip without loss of materials. He should check to assure the skip empties all material into the mixer without waste. When trucks are hauling multiple batches, it must be assured that none of the material flows from one compartment to another when the truck bed is raised. The mixer shall be regulated so that some water shall flow into the drum in advance of cement and aggregate and shall continue to flow for a specified period after all the cement and aggregates are in the drum. When using plant-mixed concrete, not more than 45 minutes should elapse from the time water is added to the mix until it is deposited on the grade when hauled in non-agitating trucks, nor more than 90 minutes when hauled in mixing or agitator trucks.

H. Limitations of Mixing

1. *Cold Weather Concreting.* Normally the mixing and placing of concrete pavement should be discontinued when the ambient temperature reaches 40°F. and is

descending and shall not resume until ambient temperature reaches 36°F. and ascending. When concrete is mixed and placed at ambient temperatures lower than those noted above, the temperature of the mixed concrete shall be not less than 50°F. and not more than 90°F. at the time of placement. When it is necessary to heat mixing water and/or aggregates, they shall be heated to not less than 70°F. nor more than 150°F. in such a manner that the mass is uniformly heated. The possible occurrence of overheated areas must be avoided. No concrete shall be placed on frozen subgrade nor shall frozen aggregates be used in the concrete.

Concrete placed during cold weather should be produced, delivered, and finished in a manner to provide for a minimum loss of heat. Finished concrete must be cured and protected in a manner that will protect it from freezing until the design strength is attained.

2. *Hot Weather Concreting*. When hot, dry, and windy conditions prevail, it may become necessary to take precautions to prevent rapid surface drying, rapid temperature changes and undesirable high temperatures (recommended maximum 85°F. to 90°F.) in the concrete during the early stages of hardening. These conditions may remove moisture from the pavement surface faster than it can be replaced by normal bleeding and cause plastic shrinkage cracks to form. It may be desirable to cool the mixing water and aggregate stockpiles to lower the temperature of the concrete. The forms may be cooled by sprinkling with water or by dragging a piece of wet burlap over them immediately ahead of concrete placement. The application or placement of curing materials immediately upon completion of finishing becomes extremely important and under some conditions, it may be necessary to use wet burlap or cotton mats for the first 24 hours. The wet burlap or mats can then be used for the remaining curing period or removed and replaced with other curing materials.

3. *Protection in Case of Rain*. Prior to start of paving operations, the Inspector should be assured that the Contractor has sufficient material on hand, such as burlap, polyethylene sheeting, or other approved material, to properly protect the pavement surface in case of rain. Sudden showers which might occur during paving operations or immediately after finishing operations require the exposed surface of the fresh concrete to be covered to prevent washing cement from the surface. Mixing and placing of concrete should cease immediately in the event of rain. If rain continues only for a short period, the protective covering may be removed and finishing completed. In case rain continues, finishing may be accomplished by rolling back a few feet of the protective cover at a time and replacing it immediately after finishing is done. Pavement surface must be inspected as soon as possible to determine the extent of damage, and the Contractor advised immediately of any corrective action or removal necessary.

I. Placing and Consolidating Concrete

1. *Slip-Form Paving.* When using slip-form equipment, the uniform distribution of batches is as important as for formed pavement. For the purpose of metering the correct quantity of concrete for the full paving width to the main screed, pavers of this type are normally equipped with an initial strike-off blade provided with power travel fore and aft independent of the forward travel of the paver. Because of the relatively large screed area and because of physical limitations as to the weights of the machine, the importance of using concrete of proper consistency and uniform distribution is extremely critical. Piles of dry concrete will cause the paver to "float" or lift above the true grade and result in a high area or bump. Consolidation of concrete in slip-form paving is accomplished by means of vibrators working within the mass of concrete. It is critical for smoothness of pavement surface that the excess concrete in advance of the forward screeds necessary to fill low spots be small enough to roll along the surface rather than be pushed or shoved. Refer to the Guide Specifications for further instructions.

2. *Fixed-Form Paving.* Formed pavement concrete should be placed on the subgrade in such a manner as to prevent segregation and to require a minimum of redistribution. Paving mixer buckets are best discharged while still in motion, moving away from the paver. Concrete dumped in piles causes non-uniformity of consolidation and additional strain on the forms and spreader. This in turn will cause increased finishing work to obtain a good riding surface. When it is necessary to spread concrete by hand, it should be done with a shovel instead of a rake or similar tool. Extreme caution must be employed when placing concrete around joint assemblies, dowels, expansion joints, etc., to avoid displacement. The use of front end loaders and other equipment riding on the subgrade is not acceptable. Every precaution must be taken to avoid disturbing the subgrade since depressions or other variations will be reflected in the finished concrete surface. Refer to the Guide Specifications for further instructions.

J. Test Specimens. See Subsection 501.02 of this manual.

K. Strike-Off of Concrete and Placement of Reinforcing Steel. Concrete is normally spread with a blade-type or screw-type spreader. The spreader should be adjusted so that concrete is struck off uniformly across the entire area. When distributed reinforcement is used, it may be placed by striking off the concrete at the proper elevation and placing by hand methods, by the use of an approved placing machine vibrating the reinforcement into proper position after full thickness of concrete is placed, or by securing the reinforcement in proper position before placing concrete. When distributed reinforcement is placed by striking off the concrete, not more than 30 minutes shall elapse before placing the top layer of concrete. If an approved placing machine is used, it should be checked for proper adjustment to assure reinforcement is at the proper final elevation and does not crawl during placement. Continued checking of location of reinforcement should be made regardless of the method of installation. The final pass of the spreader should leave concrete slightly high to provide material for finishing. When hand methods for spreading are permitted, a template, mechanical or otherwise, should be used to assure that it has been struck off to the desired elevation.

When wire mesh, bar mat, or continuous reinforcement is required, it must be placed at the proper location by hand or mechanical methods. It should be stored in a manner to avoid damage and must be kept clean.

All dowels and reinforcing should be free of dirt, oil, paint, grease, and excessive rust when concrete is placed.

L. Joints

1. Dowel-Supporting Assemblies. When dowel-supporting assemblies are required for transverse joints, they should be laid out and marked in such a manner that the exact centerline of the assembly can be reestablished. Dowel bars should be secured in place by approved supporting assemblies before concrete placement or implanted in fresh concrete by approved mechanical equipment. The tolerances of placement and type of bar should be in accordance with the applicable specifications. Where mechanical implementation is utilized, it is critical that the placement tolerances be checked on a sampling basis by carefully removing the cover of concrete. The frequency of such checking depends on the uniformity of results obtained. Small wires used for holding dowel-supporting assemblies together during fabrication and shipment should be cut after installation. Dowels should be free-moving and coated with approved lubricant as prescribed by the AASHTO Guide Specifications.

2. Tiebars. Tiebars installed across the centerline should be parallel to the surface and at approximate right angles to the centerline. Unless an approved mechanical device operating immediately behind the spreader, or behind the strike-off for slip-form paving is used for the installation, they should be installed ahead of placing the concrete and held securely in position.

3. Expansion Joints. Expansion joints should be placed at locations and by methods shown on the plans and as required by the specifications.

4. Keyways. Keyways for multiple lane paving must be held in proper position against the face of the roadway forms. Tiebars or hook dowels must be correctly spaced and securely fastened.

5. Sawing. When sawing is required or permitted by the contract, the Contractor should be required to have sufficient saws and blades on hand to perform sawing operations at the proper time. Facilities for night sawing should also be available. The time for sawing may vary from 4 to 48 hours, dependent on joint spacing, weather and atmospheric conditions. Joints should be sawed to a minimum depth of 1/4 the thickness of the pavement in a progressive manner and as soon as possible to prevent cracking without excessive raveling. Sawed transverse joints generally have a minimum width of 1/4 inch and longitudinal joints 1/8 inch. Slight raveling is not objectionable and generally is an indication that sawing is being done at the proper time. Joints must be sawed over the exact center of load transfer devices and perpendicular to the surface. When a crack occurs ahead of a saw cut, sawing on the joint should be stopped immediately and saw moved ahead several joints. A joint should be sawed at this location, then return and cut the intervening joints. Measurements of the depth and width of sawed joints should be made periodically to determine compliance with the requirements and the measurements recorded as a part of the permanent record. The shape factor should be achieved in accordance with the plan details.

6. Sealing Joints. Joints, sawed or formed, should be clean and surface dry at the time of sealing. Sealing of joints with approved material should be done prior to opening to any traffic. If heated joint sealing material is used, it should be stirred to avoid localized overheating and the temperature continually checked to assure compliance with the manufacturer's recommended temperatures. Pouring of joints should be done in such a manner that the material will not be spilled on the exposed surface of the concrete. Poured joint-sealing material should not be placed when the air temperature in the shade is less than 50°F. unless approved by the Project Engineer. Refer to the specifications for detailed instructions regarding the use of elastomeric gaskets.

M. Final Strike-Off, Consolidation, and Finishing. Immediately after spreading, the concrete should be screeded and consolidated by the use of an approved finishing machine or other approved equipment. The finishing machine should have at least two oscillating-type transverse screeds in proper adjustment. The purpose of this machine is to assist in consolidating the concrete and to leave the surface with a uniform texture and to a reasonably correct elevation and cross section for final finishing. When the spreader and the finishing machine are properly adjusted, there should be a uniform roll of concrete in front of each screed. The roll in front of the first screed should be larger than the roll in front of the back screed. The roll in front of the back screed should be sufficient to provide a uniform surface while leaving enough material for final finishing. If an excess of concrete is being carried, it will tend to lift the screeds off the forms. In addition, there will be surging behind the screed resulting in overloading of the following equipment. As the work progresses, the tilt and speed of the screeds may require adjustment to compact the particular mix being used, to eliminate tearing, and to control the amount of surge. With stiff, harsh mixes, the screed-oscillating speed should normally be rapid with a long stroke and slow forward speed. With more fluid mixes, the screed action should be decreased, both in speed and length of stroke with the forward speed increased. The number of screedings will be determined by field conditions. Excess screeding should be avoided since it tends to result in undesirable quantities of low-strength mortar on the surface.

Vibrators, which are normally attached to the back of the spreader, the front of the finishing machine, or on a separate barrier, may be pan type or the internal type with either immersed tube or multiple spuds. They are to be mounted in such a manner that they will not come in contact with reinforcement, subgrade, or forms. The entire width of the pavement must be thoroughly vibrated in a manner that will be effective for the full depth. Concrete should be thoroughly vibrated along the edges and expansion and key joints. When equipment, other than vibrators, is used for consolidation, it must meet the approval of the Project Engineer and produce satisfactory results.

After consolidating and screeding, the concrete is to be floated to remove irregularities left by previous operations and by shrinkage. This operation may be done by the use of a transverse float properly adjusted as previously described. The time of floating will depend on field conditions and it is desirable that initial settlement of the concrete be complete. If the concrete has not been thoroughly compacted and is in the early stages of shrinkage when the float passes, the final surface may eventually be rough. Floating should be held to a minimum during the period of greatest bleeding since working the surface in the presence of excess water leaches out a portion of the cement and produces low-strength surface mortar. Excessive floating should be avoided. If cutting or filling is required, all paving equipment should be checked and necessary adjustments made to eliminate the condition.

When the transverse float is used, the time of operation must be adjusted to field conditions. The screed or screeds working ahead of the transverse float should carry a uniform roll of concrete so that the transverse float will leave a smooth uniform surface free of screed marks with a minimum of surging.

Regardless of the type of float used, a continuous operation at a uniform rate of speed is necessary for obtaining the most desirable finished product.

Additional water added to the surface to facilitate finishing should be avoided. However, if necessary, it should be used sparingly and applied only with a fog spray.

When hand methods of finishing are permitted, the surface should be floated with a hand-operated longitudinal float of specified size. It should be straight and of a rigidity to prevent flexing or warping. The float is operated from foot bridges spanning the entire width, worked with a sawing motion while being held parallel to centerline and worked from one side of the pavement to the other. Each pass should overlap the preceding pass by at least one-half the length of the float.

When it is necessary to smooth or fill in open-textured areas in the pavement surface after the preceding floating, it will be permissible to use an approved long-handled float. Care must be exercised in this operation to avoid distorting the surface. The use of this equipment should be limited to small areas and should not be used to float the entire surface.

After all floating has been completed, any excess water or laitance should be removed from the surface of the pavement by using a straightedge 10-feet or more in length. Successive drags are to be lapped one-half the length of the blade. The surface should then be tested for trueness in the prescribed manner with a 10-foot straightedge that has been checked against a master straightedge. If high or low spots are observed, concrete should be added or removed and the area refinished and checked. The straightedge used for testing the surface should not be used for finishing or moving of concrete. Checking of the surface must continue until it conforms to grade and cross section and is free of irregularities.

After the floating has been completed and excess water removed, but while the concrete is still plastic, depressions should be filled immediately with freshly mixed concrete, struck off, consolidated, and refinished. High areas should be cut down and refinished. Special attention should be given to assure that the surface across joints meets the requirements for smoothness. Surface corrections shall continue until the entire surface is found to be free from observable departures and the slab conforms to the required grade and cross section.

The final finish of the surface is normally accomplished by one or more methods. The most widely used are broom, belt drag, or other texturing device. When a broom is used, it shall be of the push-broom type, generally not less than 18 inches in width and made of good quality bass or bassine fiber not more than 5 inches in length and be so operated as to produce corrugations uniform in appearance not more than 1/16 inch in depth. The most satisfactory finish is obtained when the water has practically disappeared. Finishing belts should be of an approved type, operated with short transverse strokes combined with rapid advance. This should be done when water sheen has practically disappeared and just before concrete becomes non-plastic. Drag finish may be accomplished by the use of a seamless strip of damp burlap or cotton fabric dragged longitudinally along the full width of pavement. The drag should be maintained clean and free of encrusted mortar and produce a surface uniform in appearance. Refer to the AASHTO Guide Specifications for detailed instructions regarding tine, turf, and other texture finishes. For tine finishes, the longitudinal axis of the tines should be at an angle of 10°-15° with the concrete surface to minimize dragging mortar.

The surface should be textured either by broom, belt, drag-finishing metal tines, or equipment which creates transverse grooves in the plastic concrete as soon as the water sheen has practically disappeared. All texturing shall be performed as specified in Subsection 501.03 (m.) of the AASHTO Guide Specifications.

Edging of the pavement surface adjacent to forms, joints, etc., should be done with an edger having a radius of about 3/8 inch. The concrete should have set enough to permit the edges to hold their shape after they have been finished. Care must be exercised to assure that the leg between the concrete and form or joint is held vertical. Marks left on the pavement surface by edging may be removed by a wet paint brush or a small piece of damp burlap.

If the specifications permit, instead of sawing the longitudinal contraction joints, the Contractor may place a plastic strip to create a weakened plane. A rigid, non-elastic plastic material should be used. Polyethelene is elastic. Polyester is non-stretching and non-elastic. Great care should be exercised to make sure that the strip is installed in a near vertical position. A tipped joint will result in early spalling of the concrete adjacent to the joint. The strip shall be placed by means of a mechanical installation device which will vibrate the plastic concrete sufficiently to cause an even flow of concrete about the joint material. The concrete should be free of segregation, rock pockets, or voids. The strip in the longitudinal joint should be not more than 0.01 feet below the surface of the concrete except near the crown point or to provide for plastic concrete texturing.

A construction joint shall be made at the end of each day's paving by placing a header board transversely across the pavement. Uncapped dowel bars shall be installed in the joint, seeing that the dowels are parallel with the centerline and profile of the pavement. The ends of the dowels projecting from the header should be protected so that they will not be disturbed or moved from the correct positions.

In transition areas between crowned and superelevated pavement or at the start of paving operations or for continual monitoring purposes, it may be desirable to have an independent check on the accuracy of the adjustments to the autograder and paving equipment by sampling the depth of concrete pavement placed.

N. Surface Testing. As soon as concrete has set, the surface should be checked with a straightedge and if the specifications require it, a profilograph shall be run on the finished pavement.

The Contractor is not responsible for pavement placed by another contractor or for an existing bridge or approach slabs constructed on a separate contract. When leaving or approaching such joints, the center of the profilograph will be started or stopped on the pavement to be profiled at a point approximately 15 feet from the joint. The remaining areas that are unprofiled would be checked for smoothness with the 10-foot straightedge in accordance with the current practices used on bridge decks.

Since the primary goal is to obtain a smooth pavement, it is advisable to run the profilograph over the joints at the beginning and end of the project as well as any intermediate joints as described above and exclude these readings from the profile index. Should these areas meet straightedge tolerances but not that for the profilograph, . . . then consideration should be given to grinding which would be performed at agency expense.

The specifications require that a daily profile index be determined for each day's paving. For the purposes of determining the "daily profile index" two or more indexes may be averaged together. The "daily profile index" may also be used to identify those areas having high points in excess of 0.4 inches which must be reduced by abrasive means until reruns indicate that the area does not exceed the allowable deviation. The longitudinal "profile index" of the pavement is based on the elevation of any point on the pavement relative to the elevation of points 12.5 feet ahead of and behind the point. This is measured by a 12-wheeled vehicle having a 25-foot wheel base and a reference wheel, free to move in a vertical direction, suspended midway between the outer wheels. The vehicle is calibrated to record longitudinal travel and vertical variations in elevation on a continuous strip chart as it traverses a section of pavement. The "profile index", which is determined from the recorded chart of each 0.1 mile section, is defined as the cumulative total of recorded elevation extremes—above or below a standard variation of +0.1 inch. The elevations area recorded in 0.05 inch increments.

For example, if the chart for a 0.1 mile section showed all elevation extremes to be within the +0.1 inch standard except for 2 points which measured +0.2 and +0.3 inch respectively, the "profile index" would be 0.3 inch per 0.1 mile, or 3 inches per mile.

The "daily profile index" may be used for acceptance purposes should the various individual indexes used to determine the "daily profile index" not exceed 1 inch per any 0.1 mile section or 10 inches per mile.

Grinding depths should be limited to 3/8″. If the specifications cannot be met with this, the section should be removed. Low areas which grinding cannot feasibly remedy shall be sandblasted, filled with epoxy bonded mortar and texturized by grinding. Areas which exhibit improperly finished surfaces and would require extensive patching should be removed at the Project Engineer's discretion.

O. Curing. Refer to specifications for detailed instructions.

P. Removing Forms. Removal of side forms requires that good judgment be exercised since weather and temperature will affect this operation. Unless otherwise permitted, forms should not be removed from freshly placed concrete until it has set for at least 12 hours. In all cases, the concrete should be hardened to the extent that spalling or other damage will not occur. Immediately upon removal of forms, all honeycomb must be patched and the pavement edges cured in an approved manner. Major honeycombed areas will be considered as defective work and must be removed and replaced. Any area or section so removed should be not less than 10 feet in length nor less than the full width of the lane involved. Any remaining portion of the slab adjacent to the joints that is less than 10 feet in length should also be removed and replaced.

Q. Repair of Defective Pavement Slabs

See this subsection of the AASHTO Guide Specifications.

R. Protection of Pavement

The edges and the surface of the pavement should be protected against any damage by the public, and Contractor's forces and equipment.

Any damaged pavement must be repaired or replaced prior to acceptance.

S. Opening to Traffic

The completed pavement should not be opened to traffic until it has attained a flexural strength of 550 pounds per square inch when tested by the three-point method or 650 pounds per square inch when tested by the mid-point method. This is determined by test beams prepared as described elsewhere in this manual.

When the strength of the pavement is determined by compressive specimen, it should have attained a strength of 3500 pounds per square inch.

If such tests are not conducted, the pavement shall not be opened to traffic until 14 days after the concrete is placed.

Prior to opening any pavement to traffic, it should be cleaned, properly signed and marked, and cleared of all obstructions to make it safe for the traveling public.

T. Tolerance in Pavement Thickness

Prior to final acceptance, the pavement should be core-drilled in the manner and at locations specified. These cores will be measured by the average caliper measurement in accordance with AASHTO T148. If pavement thickness is not within the allowable tolerance, deductions for deficient pavement or its removal and replacement shall be made in conformance with the contract requirements.

Cores removed from pavement should be stored in a suitable location so the Contractor and other interested parties may observe them and check the measurements.

501.04 — Method of Measurement

Refer to specifications for detailed instructions.

501.05 — Basis of Payment

Refer to specifications for detailed instructions.

501.06 — Inspection and Records

A. Plant Inspector

The Plant Inspector is responsible for enforcing compliance with the specifications of materials and batching operations. The duties of the Inspector will start at the time materials are being accumulated and continue until final records for the project are complete. The Inspector must be familiar with the source and type of aggregate intended for use, mix proportions, moisture content, method of determining scale weights, batching equipment, tests, and reports. The Inspector should be familiar with the manufacturer's brochures on the batching equipment to understand its operation. The Contractor will normally have this information. If not, it may be obtained from the manufacturer's representative.

A field laboratory should be furnished in accordance with the specifications. This laboratory should be in a location which permits the maximum number of operations to be observed while performing the necessary functions in the laboratory. It should be maintained orderly and clean for the most effective work.

The importance of proper plant inspection cannot be overemphasized since proper proportioning of materials is one of the major steps in obtaining a satisfactory pavement. The Inspector should be guided by the concept that the entire paving project is a line operation, from raw material source to the finished slab. The quality and volume of the end product are equally dependent on each step along the line, and no amount of extra effort at one step can compensate for errors, omissions, or inefficiencies at some other step.

Prior to the start of batching operations the Plant Inspector should be assured that all equipment is of an approved design and complies with the requirements of the specifications. Hoppers or bins should be set level and loaded for at least 24 hours prior to calibration. The Contractor should have available at the proportioning plant, at all times, not less than ten 50-pound weights for calibration and verification of scales. A cradle or test platform should be provided for each scale for testing purposes.

The specific duties of the Plant Inspector should include the following:

1. Observation of the stockpiling and handling of materials to assure compliance with the specifications. Maintain inspection report records to verify that all material is acceptable prior to incorporating into work. Make a periodic check (at least daily) of the quantity of cement actually used by comparing the total quantity received to the theoretical quantity used taking into account the cement remaining on hand.
2. Be familiar with the physical characteristics of aggregates, design mix proportions, the method of determining batch quantities, scale weights, yield, effective water, cement factor, and the procedures for adjusting proportions and yield when using air entrainment. The design mix as established and approved should be included in the permanent records of the project.

3. Calibration of weighing and measuring devices, balancing scales and checking calibrations, and systematic and regular checking of scale settings for batches to assure proper quantities are being dispensed. Scale settings are not to be made by the Inspector since this is the responsibility of the Contractor's representative. Check to ensure that all scales have been certified.
4. Make free moisture tests (AASHTO T142). These tests are required so that batch weights may be adjusted and to ensure that the maximum allowable water-cement ratio is not being exceeded. The batcher should be advised immediately of any change in moisture content of the aggregates, and of the maximum gallons of water that may be added at the mixer without exceeding specification and design limits.
5. Check batch trucks used for transporting unmixed batches to the paver to assure they conform with specification requirements and that the compartments are of sufficient size, are constructed in a manner to prevent loss of material and spillage or contamination from one compartment to the other, and that the total batch is discharged. Examine batches in which the cement is in contact with the aggregates in excess of 1-1/2 hours and if lumps or a crust of hardened cement is found, discard the entire batch.
6. Require that central-mix concrete be hauled in vehicles meeting specification requirements and in a manner to avoid segregation and be delivered at the site with proper consistency and workability before the concrete starts to take its initial set. Require agitating-type trucks if these conditions cannot be met.
7. Check truck mixers to ensure that they contain a water-metering device of prescribed accuracy, approved revolution counters, mixing blades inside the drum which are not caked with mortar, loose, broken, bent, scalloped, worn 20 percent of any dimension, or otherwise damaged. Truck mixers should not be permitted to mix batches having volumes greater than the maximum capacity indicated on the manufacturer's rating plate.
8. Proper handling, curing, and breaking of beams to determine flexural strength.
9. All required records and reports should be kept current and prepared in conformance with instructions. All test reports should be dated and signed by person making the test. An orderly record should be kept of all specific checks or tests made to determine compliance with specifications such as:

   All calibrations, verifications, and checks
   Scale certifications
   Accumulative total of all acceptable materials received
   Moisture tests
   Scale weight settings
   Accumulative total of material used
   Flexural test results
   Diary:

   A diary must be maintained by the Plant Inspector and should be a concise record of daily events and observations of the Inspector. This record should include: all special instructions to the Contractor, unusual action taken by the Inspector, daily time of starting and ending operations and extent of progress by stations, lost time due to breakdown or other reasons, Contractor's forces, weather, and temperature conditions.

B. Slab Inspector

The Slab Inspector is assigned the responsibility for obtaining the construction of structurally sound pavement with the desired riding qualities, and of performing the necessary checks and tests on this portion of the operation to determine that it meets the contract requirements. To do this, the Inspector must have a thorough knowledge of the plans and specifications, recognize good construction practices, have a working knowledge of equipment used, be able to issue clear-cut decisions to the Contractor's representative, and be fully aware of all required tests and reports. It is recommended that brochures from manufacturers of the equipment to be used be made available to the Inspector. The Contractor will normally have this information. If not, it can be obtained from the manufacturer's representative.

All required records and reports should be kept current and prepared in conformance with instructions. All tests reports should be dated and signed by the person making the tests. An orderly record should be kept of all specific checks or tests made to determine compliance with specifications, such as:

Subgrade measurements (see subgrade)
Check of mixer timer
Check of impulses or strokes of vibrators or tampers
Record of water used at on-site mixer
Record of equipment checks and adjustments
Slump and entrained air tests
Depth of reinforcement
Check of finished crown and straightedging
Check of curing application
Check of depth and width of sawed joints
Diary:

A diary must be maintained by the Slab Inspector and should be a concise record of daily events and observations of the Inspector. This record should include: all special instructions to the Contractor, unusual actions taken by the Inspector, daily time of starting and ending operations and extent of progress by stations, lost time due to breakdown or other reasons, Contractor's forces, weather, and temperature conditions.

Section 550 — Concrete Pavement Rehabilitation

Section 551 — General

551.01 — Description

The effects of heavy vehicles, high traffic volumes, presence of water, inadequate subgrade or base, or corrosion of load transfer devices manifest themselves as deterioration of portland cement concrete pavement with various forms of damage. This damage will appear as cracking (transverse, diagonal, or longitudinal), joint spalling, D-cracking, corner breaks, or joint faulting. The rehabilitation or restoration of concrete pavement begins with a survey of the pavement to identify the damaged or distressed areas and to identify the severity of the damage and the underlying causes. The survey is followed by the design and then the construction of the required restoration.

The restoration procedure is composed of one or more of the following: subsealing and pavement jacking, partial depth patching, full depth patching, establishing load transfer, installing underdrains, installing pressure relief joint, diamond grinding, or joint/crack sealing.

551.02 — Materials

The materials used in pavement restoration are the materials commonly used in the initial construction of concrete pavement, except the addition of accelerating and water-reducing admixtures are usually necessary. There may also be a need for special fast-setting chemical concretes. The requirements for all the materials to be used are specified in the contract.

Section 552 — Concrete Pavement Jacking

Jacking is required when the voids under the pavement have enlarged to the extent severe pavement faulting or settlement has occurred. The pavement profile or cross-slope will be restored with jacking. This may be performed in lieu of full depth patching, providing the pavement is not badly cracked or deteriorated.

When the contract requires raising the pavement, the material and equipment are the same as used for subsealing.

The drill hole layout will be different than subsealing and will be shown on the plans. If the hole pattern is not specified, experimentation based on past experience will be necessary. The jacking should begin with the holes located at the lowest point of the dip or settlement.

When a treated base supports the pavement, it is necessary to drill the grouting hole through the pavement and the base. The hose nozzle should not protrude beneath the pavement which permits the grout to fill voids between the base and the pavement and below the base.

The grout mix can be less fluid for jacking than subsealing, especially when pumping in a hole the second or third time.

Pavement jacking is different than subsealing in that the grout injection is continued until the pavement is raised, in small increments, producing an improved surface profile. Constant attention must be given to controlling the movement of the pavement. The amount of movement can be observed with an engineer's level, stringline, or Benkleman Beam.

The hose nozzle pressure becomes more important in pavement jacking. More pressure will be required to raise the pavement than to subseal the voids. However, too much pressure resulting in rapid pavement movement can cause cracking radially from the hole, or unnecessary raising of the shoulder, blowouts outside the shoulder, or infiltration of existing underdrains.

Section 553 — Subsealing and Stabilization

Joint faulting is the primary evidence of subsurface consolidation or displacement. When the presence of voids under the pavement is suspected or has been determined to exist but pavement faulting or settlement has not developed, subsealing will be necessary to fill these voids. When subsealing, the pavement must not be raised; the grout injection is immediately discontinued when movement is detected.

Pozzolan (fly ash/cement) appears to be the most successful blend of grout that is incompressible, insoluble, and noneroding. Depending on the locality and availability, other blends using sand, silty loam, limestone dust, or bituminous material may be used although each of these have some objectionable characteristics. Selection of the grout blend used must be based on the advantages and disadvantages of locality considerations and material availability.

The required equipment is a grout mixer, grout pump, air compressor and drills, water tanker, and pavement movement detection instruments along with hoses, valves, gauges, wood plugs, and service trucks. The mixer should be a high-speed colloidal mixer (800 to 2000 RPM) or a paddle-type mixer as specified and the pump should be a positive displacement model developing 5 to 200 psi.

When a treated base supports the pavement, drilling through the pavement, but not through the treated base will be necessary which permits the grout to fill voids between the base and the pavement. For untreated bases multiple grout operations may be necessary, dependent upon viscosity of the grout and character of the untreated base material.

The drill hole spacing pattern, as shown on the plans, should be checked to verify an optimum sealing effort. When the spacing is not specified, field experimentation at the beginning of the project to determine the pattern which provides the best subsealing results will be necessary. Drill operation and drill speed must be closely monitored to avoid spalling out of the drill hole on the underside of the pavement.

Constant attention must be given during subsealing to prevent movement of the pavement. Movement can be detected with an engineer's level, stringline, or

Benkleman Beam. Grouting must cease immediately at the hole if movement is detected.

Section 554 — Resealing of Joints and Cracks

One of the final steps in rehabilitating portland cement concrete pavement is the sealing or resealing of joints and cracks.

Where pavement subsealing and/or pavement jacking has been performed, it is critical that all joints, transverse and longitudinal, be maintained in a sealed condition to avoid damage by the intrusion of surface water into the joints and subsequent deterioration of the grouting.

The intrusion of moisture and incompressibles into a joint leads to pumping, softening of subgrade, freeze-thaw action, corrosion of dowels, and spalling. Maintaining a proper seal is the best preventive step taken to ensure a longer life for the pavement.

A condition survey will determine when a pavement needs resealing. Rehabilitation work will require resealing.

All larger spalls, distressed concrete, and moving slabs must be corrected prior to sealing for the sealant to properly function.

All working joints or cracks must be prepared by grouting to remove existing sealant and incompressibles. The opening may need to be widened by diamond blade sawing to provide an improved depth/width shape factor. The specified shape factor will be shown on the plans. Clean sidewalls should be provided by sandblasting.

Any working cracks should be sawed to produce the proper shape factor. Working cracks should be identified prior to the sawing operation.

The specified sealant will be either a hot-poured elastomeric material, cold-poured silicone, or a preformed compression seal. Other types of the material that have proven successful may be specified by the Department.

Where cold-poured silicone sealant is utilized, consideration should be given to the manufacturer's recommendation of the ambient air temperature at which it is allowed to be poured. The silicone sealant's ability to adhere to the adjacent concrete surface is directly affected by the temperature at which it is poured as well as the length of time it is allowed to cure prior to traffic loading being placed on the concrete slabs.

Proper cleaning, joint shape factor, and application of the material cannot be overemphasized. The backer rod used to provide a bottom form for the hot- and cold-poured sealants must be installed to the correct plan depth. It must be compressible, resistant to heat, nonreactive with the sealant, nonshrinking, and nonabsorptive.

Preformed compressible seals must be installed to the correct plan depth. They must not be twisted or stretched. The lubricant adhesive must be applied according to the manufacturer's directions.

Joints and cracks should be sealed the same day they are prepared. This prevents unnecessary intrusion of moisture, dust, and incompressibles. The sealant material surface depth below the pavement surface will be shown in the contract and must be maintained.

Traffic may be permitted on the newly sealed pavement only after the sealant has had sufficient time to cure. This time should be identified in the contract.

When diamond blade grinding will be utilized in the rehabilitating process, the joint sealing should be postponed until after the grinding. This will ensure proper depth of sealant.

Section 558 — Partial Depth Patching

Partial depth patching can be used to repair spalls, potholes, and other surface defects, along with other distresses. Minimum patching depth should be 1/3 the pavement depth unless experience indicates a greater depth patch is needed. Spalls and other defects which extend below the midpoint and cracks which extend full depth of the pavement cannot be effectively repaired with the partial depth method.

When the contract also provides for diamond profile grinding and subsealing, partial depth patching should be made prior to the grinding but after the subsealing.

The saw cut around the perimeter of all patches is extremely important and should be as deep as the proposed patch unless specified otherwise. Sawing beyond the patch limits and across pavement joints should be sealed with epoxy cement.

A relatively lightweight pneumatic hammer is used to remove the existing concrete in a partial depth patch to prevent damaging the remaining concrete. Milling machines may be economically practical when the patched areas are large enough. The bottom of the patched area should be approximately parallel with the pavement surface.

Any patches placed adjacent to a joint or a working crack must have a means of separating the newly placed material from the existing concrete. This should be a material that can absorb any expansion. If the separating material is not of adequate width or depth to provide the proper reservoir for sealant, the joint or crack must be sawed before sealing.

All loose concrete particles, dust of fracture, oil, etc., must be removed with sand blasting and/or high pressure air.

A bonding agent must be applied to the bottom of the partial depth patched areas. The vertical faces of these patches not adjacent to joints or working cracks should also be treated with the bonding agent. The bonding agent will be a cement-sand mixture or a chemical adhesive as specified. The bonding agent must be scrubbed onto the area to be patched to thoroughly coat the entire exposed surface. The maximum elapsed time from bonding agent application to placement of the patch material should be in accordance with the manufacturer's recommendations.

Partial depth patches may use normal set concrete, accelerated set concrete, polymer concrete, quick set materials, or epoxy concrete. The time restraints in maintaining traffic and experience history with the materials will influence the type of patching material used. The selected material must not be placed until the approval of ingredients and proportioning has been received from the materials laboratory.

All patches with portland cement concrete must be densified with vibrators until all entrapped air bubbles cease rising to the surface.

Final screeding and finishing should produce a cross section matching the existing pavement or as called for on the plans. Texturing should also match the existing pavement.

Check for damage to the shoulder area by the concrete removal process or the concrete-placing equipment. This damage must be repaired or restored with similar shoulder material. The pavement-shoulder interface should be tacked or sealed.

Curing will normally be according to the manufacturer's recommendation for the patch material or in accordance with the specifications. When early traffic opening is required or ambient temperatures delay setting, plastic-treated burlap, or other type blanket material, should be secured around the pavement slab.

Section 559 — Full Depth Patching

Full depth patching will be used to repair high severity cracking (transverse, diagonal, or longitudinal); joint spalling when the spalling extends below mid-depth of the slab; D-cracking, corner breaks, and joint faulting when the condition cannot be remedied with subsealing and partial depth patching or jacking.

When the contract provides for diamond profile grinding, the full depth patching should be performed before the grinding.

The areas to be patched will be shown on the plans. The areas should be the full width of the lane and may be the full length between transverse joints, any section between the joints, or a section centered around a transverse joint. The minimum length of patch should be 10 ft.

The existing subbase should be investigated for suitability and recompacted. In some cases, it may be necessary to remove the contaminated material and replace it with material compatible with existing subbase material or increase the pavement depth in the patching area. If the pavement depth is increased, the specifications should be reviewed for required bond breaking material at the normal pavement depth.

The patches may be removed by the chip-and-break method or by the lift-out method. The lift-out method is recommended. All patches should have a full depth saw cut around the perimeter regardless of the method of removal.

When the chip-and-break method is used, the downward force of the air tool or drop hammer should be controlled to prevent damage to the existing base and spalling of the underside of the pavement. Damage to the adjacent slab should be avoided. When the existing base is disturbed, all loose material should be removed and replaced with full depth pavement patching material.

With the lift-out method, the deteriorated pavement is sawed full depth into portions that can be lifted easily with available equipment. One of the transverses direction saw cuts may have to be double-sawed in such a way to provide a relief joint which aids in reducing the wedging action. Holes that will accommodate lifting devices must be drilled through the slab. The lift-out method minimizes disturbing the existing base.

Full depth patches may use normal set concrete or accelerated set concrete. Polymer concrete or quick set materials are not economical for patches of this size. The selected material must not be placed until the approval of ingredients and proportioning has been received from the materials laboratory.

The patches may require load transfer devices to be installed. Mesh reinforcement may be required also. The mesh size and location will be shown on the plans.

Any patches placed adjacent to a joint or a working crack must have a means of separating the newly placed material from the existing concrete. This should be a material that can absorb any expansion and be easily removed without damaging the joint face. If the separating material is not of adequate width or depth to provide the proper reservoir for sealant, the joint or crack must be sawed before sealing.

All patches of portland cement concrete must be densified with vibrators until all entrapped air bubbles cease rising to the surface.

Final screeding and finishing should produce a cross section matching the existing pavement or as called for on the plans. Texturing should also match the existing pavement.

Check for damage to the shoulder area by the concrete removal process or the concrete-placing equipment. This damage must be repaired or restored with similar shoulder material. The pavement shoulder interface should be tacked or sealed.

Curing will be in accordance with the specifications. When early traffic opening is required or ambient temperatures delay setting, plastic-treated burlap, black polyethylene, or other type blanket material should be secured around the pavement slab.

Care must be taken to maintain the joint width to accommodate expansion of the pavement and to permit placing the joint sealant.

Section 560 — Diamond Grooving of Concrete Pavement

Grooving is a possible solution when wet weather accidents are a problem. Grooving provides escape channels for water and thus reduces the hydroplaning potential.

Longitudinal grooving does improve skid resistance when tested for peak friction using a bald tire. It does not change significantly when tests are made with the lock-up trailer with treaded tire.

Grooving equipment utilizes a diamond blade cutting head with larger spaces than are used on a grinding head.

Section 561 — Diamond Grinding of Concrete Pavement

High volume and heavy vehicles contribute to joint faulting and wheel rutting. Subgrade settlement causes pavement warping and disrupted transverse surface drainage.

After faulted pavement is subsealed, the riding quality and texture can be restored by grinding with diamond-impregnated blades. This can also restore the drainage. Diamond grinding may also be performed on areas that have been full or partial depth patched to achieve a uniform riding surface. All rehabilitation to be performed, except joint sealing, should be accomplished before the grinding.

Water used to cool the blades will also help to control dust. Water that is vacuumed from the surface must be disposed of. This slurry should be hauled away. . . for disposition.

The noise level should be held to a tolerable level. The project specifications may require a maximum dba threshold.

The pavement condition survey will determine the areas to be ground. These will be detailed on the plans. Grinding is usually required only in the outside or heavy traffic lanes. The inside or passing lane may have a satisfactory profile due to lower traffic volume and lightweight vehicles.

The grinding machine should be operated opposite the direction of traffic for the best results. Support equipment should be oriented in the direction of traffic which will necessitate backing up during the grinding operation. This reduces the confusion of motorists traveling in the adjoining lanes. It also permits the traffic cones to be removed behind the operation, opening the repaired lane to traffic. Grinding should progress from the low side of the cross slope to the high side to avoid ponding of water at the shoulder or longitudinal joint. If grinding is only done on one lane of multilane facilities, daylighting of the grinding to assure unimpeded cross drainage will be necessary. Traffic striping removed during grinding shall be immediately replaced.

The spacing between grinding blades can be varied according to the aggregate hardness to provide the proper longitudinal corduroy effect.

Acceptance of the grinding is based on pavement smoothness. The smoothness must be accurately measured with the device specified and will be documented to provide the smoothness desired by the agency and to be equitable in making payment to the Contractor.

Section 562 — Pavement Milling

Pavement milling is generally done in conjunction with a bonded overlay project to correct faulting, surface irregularities and to provide a suitable bonding surface for the overlay.

It is accomplished through the use of a cold milling machine that utilizes carbide tipped teeth mounted on a revolving drum striking the pavement to chip away material. Milling machines vary in width from 2 to 12 feet. The carbide tips must be continually maintained (replaced) to provide a uniform texture with no ridges or low spots. All rehabilitation to be performed, except joint sealing should be done before milling is done.

Section 563 — Portland Cement Concrete Bonded Overlays

Fully bonded PCC overlays should be used only when the existing pavement is in good condition or where all serious distress has been corrected. To achieve a fully bonded PCC overlay it is necessary to carefully prepare and clean the surface of the existing pavement prior to placing the overlay. Surface preparation includes removal of a thin layer of the existing pavement surface, and then secondary cleaning to remove remaining residue. These procedures should be capable of removing surface contaminants, paint, and all unsound concrete. Initial removal is usually done by shotblasting or cold milling.

In addition to preparing the existing pavement surface, a grout made from cement:water (neat cement) or sand:cement:water can be placed on the cleaned dry surface just in front of the paver. Typically the grout is squeegeed or broomed onto the surface, or applied with a high-pressure sprayer.

Bonding grout must not be allowed to dry or set prior to placement of the concrete for the overlay. Also the grout should be applied only to *completely* dry pavement surfaces. Where shear tests indicate adequate bond strength development will occur without the grout, the grout application can be removed from the procedure.

Bonded overlays should not be placed during the times of high daily temperature changes (e.g. early spring or late fall). If placed during these periods, extreme care in curing must be used. If not, the overlay may experience delamination problems.

Joints in the overlay should be sawed directly above joints in the existing slab as soon as it is deemed feasible. To avoid secondary joint cracking, transverse joints should be cut completely through the overlay where the overlay is less than 4 inches in thickness. Longitudinal joints should be sawed to one-half the nominal overlay thickness.

Section 564 — Portland Cement Concrete Unbonded Overlays

Unbonded concrete overlays are achieved through the provision of a debonding interlayer between the existing and overlay layers. The debonding interlayer will isolate the old and new layers, making this technique very effective for considerably distressed concrete pavements.

To effectively retard or arrest the reflection of distress from the underlying pavement, the thickness of the bond breaker material used in the interlayer must be sufficient to allow for completely independent action of the slabs. Asphalt concrete of one-inch nominal thickness is typically used for this purpose. Thinner interlayers can also be successfully used where the existing pavement distress is not as severe.

Unbonded overlays should not be placed on bituminous interlayer materials which are excessively hot and soft. This can occur when the sun's heat is soaked up by the black material during hot summer days. To prevent this from occurring, a whitewash of either lime slurry or curing compound can be used to paint the surface white and consequently reflect the heat. A thin fogging of water can also be used for this purpose.

Good load transfer from an effective sleeper slab arrangement can be obtained by mis-matching the overlay joints, by at least three feet, from joints in the underlying pavement. Mis-matching will help ensure good load transfer is maintained at the overlay joints over the life of the overlay. It may be desirable to match expansion joints due to their excessive movement.

Section 565 — Portland Cement Concrete Direct Partially Bonded Overlays

Partially bonded PCC overlays result whenever fresh concrete is placed directly on relatively sound, clean existing slabs. No attempt is made to either bond the overlay or prevent bond.

As with bonded overlays, it is very important that all structural defects be corrected in the existing pavement before the overlay is placed. This will provide the overlay with uniform support and prevent "keying" of the overlay into defects in the existing pavement.

Transverse and longitudinal joints in a partially bonded PCC overlay should always be sawed directly above joints in the existing pavement and should be sawed as soon as it is deemed feasible. The type of joint should also match that of the existing pavement. In other words expansion joints should be placed above existing expansion joints and contraction joints above existing contraction joints, etc. The depth of sawcut for both transverse and longitudinal joints should be one-third the nominal overlay thickness.

Section 566 — Portland Cement Concrete Pavement Recycling

Portland cement concrete pavement recycling is used when an existing PCC pavement is too extensively damaged to be rehabilitated by other methods. It is generally a good source of quality aggregate because of the high quality aggregate normally placed in PCC pavements. Once crushed and properly sized, the material can be used as standard aggregate in the production of Portland cement concrete.

Specifications usually restrict, or in some cases do not permit, the use of recycled fines in a concrete mixture. This is because crushed fines have a high absorption and high angularity which tends to inhibit workability. Where restricted a maximum of 30 percent recycled fines are allowed in the fine aggregate portion of the mix. As a rule of thumb, a recycling operation generally produces 1.5 times the coarse aggregate required to replace the pavement to the same width and thickness. Therefore recycling can supply all or most of the coarse aggregate required for increases in pavement thickness, widened lanes and/or concrete shoulders.

On a recycling project, the existing pavement must be processed, on site, into slab sizes that are readily handled and transportable to the crushing operation. A 2-ft. to 4-ft. square size is often used depending on the equipment available.

Numerous pieces of equipment have been used to break up the existing pavement including:

1. Diesel hammers mounted on a trailer. The hammers fracture the slab as the trailer is towed.
2. Guillotine type breaker with a drop weight of 6.2 tons.
3. High frequency, low amplitude vibratory (resonant) pavement breakers.
4. Pneumatic hammers or hydraulic chisels mounted on a backhoe or end loader. The equipment punches holes at preset intervals allowing the loader to easily break the slabs.

The most difficult problem with recycling is the removal of any existing steel. In the past during pavement removal operations, torches, or pneumatic or hydraulic shears, were provided to cut the steel that was not broken in the fracturing process in order to allow the loader to handle the slabs. This was required because older pavement breaking operations did not break the slabs as effectively as the more advanced technology provides.

On more recent projects, all steel has been removed at the crushing plant, even on long jointed (99 feet) heavy mesh-dowel pavements. A 30,000 foot-pound diesel hammer and a "rhino-horn" to lift and stack the broken concrete can provide pieces easily handled by a front-end loader.

A normal crushing plant can very easily process recycled concrete with only a few minor modifications that can be left in place to process conventional materials. Steel removal is the major operation that must be accommodated in the crushing process. With the smooth steel used in older pavements, primary crushing will separate nearly all the steel from the concrete. On occasion, the primary crusher may have to be shut down for steel removal. Dowel bars and other large pieces of deformed steel have caused problems in the primary crusher. This can usually be controlled through the use of hand picking of steel along the primary crusher conveyor, or through the use of an electromagnet to attract steel prior to entering the crushing bin.

Section 567 — Portland Cement Concrete Shoulders

Placement of tied concrete shoulders as a rehabilitation procedure should be considered when the existing shoulder is deteriorated and in need of replacement, significant distress has occurred in the outer lane, and the PCC does not have serious durability problems.

Tie bars or hookbolts must be grouted in place in holes drilled into the outer lane slab edge. If the outer lane concrete is deteriorated by reactive aggregate or "D" cracking, the tie bars will have to bond with and bear against poor material which will likely deteriorate under bearing stresses induced by edge loadings. Once deterioration occurs, the tie bars will no longer transfer load from the mainline slab and will have consequently forfeited the structural effectiveness of the shoulder.

In addition the need for sound existing concrete in which to place the tie bars, the spacing of the tie bars is critical in order to produce the desired load transfer between the mainline pavement and the new shoulder.

Typically the shoulders are placed to the same thickness as the existing mainline pavement along the outer lane edge. The shoulder thickness can be tapered thinner at the outer edge, but only across the top of the shoulder. The bottom of the shoulder slab must remain at a cross-slope equivalent to or greater than the cross-slope of the bottom of the outer lane to promote free drainage.

Transverse shoulder joints should match those in the existing pavement. Spacing of these joints should not exceed 20 feet. If the mainline pavement joint spacing is greater than 20 feet, intermediate shoulder joints should be placed. These joints do not require dowel bars or other mechanical load transfer devices.

DIVISION 600 — MISCELLANEOUS CONSTRUCTION

Section 601 — Concrete for Minor Structures and Incidental Construction

References: Guide Specifications Section 601, and applicable materials in Section 700.

The concrete for minor structures and incidental construction shall meet the specified mix requirements. The Inspector should assure that the concrete is properly mixed and meets the requirements in regard to slump, air content, uniformity, and desired workability when it is placed.

The placement of concrete should not proceed until the Inspector has approved the site, reinforcing steel, and forms to assure the details of the contract are being followed. The concrete should not be placed on an unsatisfactory subgrade. The placement of the concrete should be continuous for each pour sequence to avoid a cold joint and the concrete shall not free fall more than five (5) feet to avoid segregation of the mix by utilizing chutes, troughs, or pipes. The concrete shall be adequately vibrated to assure proper consolidation of the mix. Vibrators shall not be used to cause the concrete to flow or run into position in lieu of placing nor shall vibration be prolonged to cause segregation of the mix. The concrete should be struck-off and finished as specified. Excessive use of water to facilitate the finishing operation should be discouraged due to weakening of the upper portion of the concrete, which can lead to premature spalling or scaling.

The concrete shall be cured as specified in the contract.

Section 602 — Reinforcing Steel

References: Guide Specifications Section 602, and applicable materials in Section 700.

The Inspector should assure that the reinforcing steel is of the proper type, size, and shape specified and free of dirt, scale, detrimental rust, paint, oil, and other foreign substances.

The reinforcing steel shall be placed as specified in the contract and verified by the Inspector. The Inspector should assure the reinforcing steel has adequate fasteners and supports to securely hold the reinforcing steel in the proper position during placement of the concrete.

Concrete shall not be placed until the reinforcing steel is acceptable.

Section 603 — Culverts and Storm Drains

Reference: Guide Specifications Section 603, and applicable materials in Section 700.

603.01 — Minor Drainage Structures

Minor drainage structures are those of less than a 20-foot span, including culverts, sewers, manholes, and underdrains. Plans and designated stationing of structures must be reviewed at the respective sites for field condition aptness. Planned changes, additions, or deletions should be brought to the Contractor's attention at the preconstruction conference and proper procedures followed to implement the revisions.

Preliminary measurement data shall be recorded in the field notes for the final estimate computations.

603.02 — Excavation

Excavation for a structure should be adequate to accommodate the structure to be installed and should provide sufficient working space and room for forms and bracing if required. Economy will generally dictate that the excavation be the minimum necessary. This is usually desirable for stability also. Undisturbed natural ground under and at the sides of the structure will ordinarily furnish better support for the structure itself and the embankment to be constructed over it, than will backfill.

The trench for an underdrain should be at least 8 inches wider than the diameter of the pipe. A trench may be dug with vertical sides in stiff clay or other stable earth but in most other types of material will have to be sloped to counteract the tendency to cave in. A slope of 1:6 will ordinarily be sufficient in firm earth if the trench does not stand open too long. In less stable materials it may be necessary to excavate a 1:1 slope or to resort to the use of bracing or sheet piling to hold the sides of the trench in place. Applicable safety standards should be enforced to ensure that a safe work environment is maintained.

The excavated material should be cast well back from the edge of the excavation so it will not slough back in and cause the bank to cave.

Payment for structure excavation should be in accordance with specification requirements.

603.03 — Foundation

It is essential that the foundation under a structure provide support as firm and as nearly uniform as possible under the entire bearing surface. Bedding, as required, should be utilized. The foundation should be shaped with a template to conform to the shape of the pipe culvert or storm drain. The bottom of the excavation should be on solid ground for its full length and width. Culverts should not be placed partly on filled ground and partly on undisturbed natural ground because of the probability of unequal settlement which might distort or break the structure. This applies transversely as well as longitudinally. When a side hill location is used, the culvert should be benched into the hillside far enough to be entirely on solid ground. If the culvert must be on filled ground, the filled material should be placed in thin layers with proper moisture and thoroughly compacted to provide a foundation nearly comparable to that afforded by the natural ground.

The installation of drainage structures or systems in embankments should be avoided when practicable. This reduces the possibility of providing a foundation subject to settlement which could cause breakage of the structure or low spots which would not drain. When such an installation must be made, the embankment should be constructed and thoroughly compacted to a specified height above the elevation of the bottom of the structure. The excavation should then be made in the compacted fill.

Unstable foundation material should be removed and placed with satisfactory material. If this cannot be done reasonably, a layer of sand, gravel, or other suitable material should be placed on the foundation and worked into the unsatisfactory material until a stable foundation is formed. If a pipe culvert is to be placed in rock excavation, the rock should be removed as specified below the bottom of the pipe and a well-compacted cushion of gravel, sand, or other suitable material should then be placed and shaped to conform to the bottom of the pipe as a bed for the pipe. When bell and spigot type is used, bedding or foundation should be shaped to conform to the bells so that the pipe will have uniform bearing throughout its length.

Cambering of a culvert grade line should always be considered before installation of the pipe is started. Subsidence varies widely depending on the fill height, the depth of foundation soil to a solid stratum, and the compressible character of the foundation soil.

Camber should not be used as a substitute for foundation stabilization. Poor foundation should be corrected before installing culverts and the amount of camber should be based on the foundation soil profile after stabilization.

603.04 — Installation

After the foundation has been completed and alignment and grades have been established for the structures, make sure materials to be used have been tested and no damage has occurred to previously approved items prior to their placement. No installation should be made on frozen earthwork.

A. Pipe Culverts. Pipe laying should begin at the downstream end of the culvert with the bell or groove ends facing upstream. Care shall be taken to see that each section will have full contact with the preshaped foundation or bedding. When elliptical pipe with circular reinforcement or circular pipe with elliptical reinforcement is used, the pipe should be laid in such a position that the manufacturer's marks designating the "top" and "bottom" of the pipe shall not be more than 5 degrees from a vertical plane through the longitudinal axis of the pipe.

For reinforced concrete pipes 30 inches in diameter, or less, joints shall be finished on the inside surface before the grout or mortar in the joints has set. For pipes larger than 30 inches in diameter, the inside joints shall not be finished until the backfill over the pipe has been completed.

Corrugated metal pipe culverts shall be laid with the separate sections joined firmly together and with the outside laps of circumferential joints pointing upstream, and with longitudinal laps on the sides.

When shop strutting of pipe is called for on the plans, no additional vertical diameter elongation is required or permitted in the field.

603.05 — Structural Plate

This item is covered in Section 23 of the AASHTO *Constructional Manual for Highway Bridges and Incidental Structures*.

603.06 — Concrete Structures

When concrete is to be placed, all aggregates, cement, water, reinforcing steel, joint fillers, drain pipes, admixtures, and curing compounds must be approved prior to use. An authorized concrete design mix shall be available. Foul weather protective equipment must be at the site in accordance with specifications. Mixing equipment, including volumetric mixers, shall be capable of producing the amount of concrete required within time to avoid unplanned construction joints. Field-testing equipment for control of the mix shall be at hand.

Horizontal construction joints should be placed only as permitted by the plans or approved by the Project Engineer. Vertical construction joints may be required by the plans due to the length, or to build the structure in two or more parts to maintain traffic.

Design drawings of all concrete box culverts must be checked carefully for proper location and skew, and for most efficient angle of wingwalls.

603.07 — Forms

The forms for the concrete should be constructed accurately to the required lines, elevations, grades, and contours and sufficiently braced and tied to prevent any settlement, distortion, separation, or displacement. The forms may be of metal or wood but must be watertight and strong enough to carry the loads and stresses to which they will be subjected, including the effect of vibration if it is to be used in compacting the concrete. The forms should be so constructed that they can be removed without damage to the concrete so that sections to be removed before others can be removed without disturbing the other sections or reducing support.

The inside faces of forms which will be exposed to the concrete should be thoroughly oiled, soaked with water, or greased to prevent absorption of mixing water and adherence of mortar. Cleanout openings should be provided where necessary to permit the removal of sawdust, dirt, and other objectionable material from the forms before the concrete is placed.

Check all forms for accuracy, strength, and compliance with approved form plans. Adequacy of the forms is the Contractor's responsibility, but any inadequacies which might affect safety or the acceptable quality of the completed work should be brought to the attention of the Contractor for correction before any concrete is placed.

Check all steel to see that it will stay in its proper place during pouring, vibrating, and finishing operations.

603.08 — Removal of Forms and False Work

If finishing is required, all forms, except those under slabs, beams, girders and arches, and other supporting heavy vertical loads, should be removed as soon as the concrete has set sufficiently to remain in position without the support of the forms. The time of removal will be in accordance with the specifications.

603.09 — Curing

Proper curing is of major importance. It consists of keeping the exposed surface of the concrete moist to prevent loss of mixing water by evaporation and to protect against extremes in temperature. Curing should proceed in accordance with applicable specifications.

The method used should be checked frequently for compliance with AASHTO Guide Specifications. Any deficiencies must be immediately directed to the Contractor's attention for correction.

603.10 — Backfilling

Backfill material should be similar to the adjacent material. When this is not possible, a transition should be made to minimize differential frost action. In any event, the material should be free from muck, large stones, lumps, and rubbish. To obtain uniform pressure against the pipe or structure, the backfill material should be placed in layers about 6 inches thick and thoroughly compacted. Add water if necessary to bring the material to the optimum moisture content for maximum consolidation. To avoid displacing or unduly stressing the structure, backfill on both sides simultaneously.

Pipe culverts should be backfilled with compacted material on each side of the pipe at least equal to the diameter of the pipe, except when undisturbed material is present. The compacted backfill should extend at least 8 inches, and preferably a distance of two diameters, above the top of the pipe. For structures such as box culverts, abutments, and retaining walls, the berm of compacted material should extend behind the structure a distance at least equal to the height of the wall being backfilled or to undisturbed material. Special care should be given to tamping material under the haunches of pipes. Excessive compactive effort under the haunches may raise the pipe above intended grade.

Density tests shall be made in adequate or required number. Material with low density should be removed and replaced with material which is fully compacted.

The backfill for trenches and other small areas should be deposited and compacted in thin layers. Mechanical tampers are normally used. Whenever space will permit, the backfilling should be done by means of special backfilling attachments for tractors and power shovels, or suitable equipment. Compacting should be accomplished by means of rollers. Pipe culverts and other structures should be adequately protected from damage before any heavy equipment is operated near or over them.

The field notes for the individual structure, in addition to administrative information and preliminary-final measurements, shall include: remarks and reasons for increase or decrease of plan quantities, reasons for undercutting excavation, source of backfill and quality thereof, compaction data, torquing record, unusual difficulties of installation, and conditional acceptance dates.

Section 604 — Manholes, Inlets and Catch Basins

Reference: Guide Specifications Section 604.

Prior to the Contractor starting to work on a structure, the Inspector should examine the plans and existing drainage conditions to assure that the structure will be properly located.

All material should be inspected prior to being incorporated into the work. The Inspector should also assure that all materials have been approved for use in the work and that all the required certifications have been received. When structural concrete is used, the inspection requirements for structural concrete as mentioned in this manual should be followed. For precast concrete units, the same material inspection requirements as for reinforced concrete pipe should be enforced.

Before the Contractor begins the construction of the structure foundation, the Inspector should inspect the soil foundation. The foundation material should be firm and dry in order that the structure will be properly supported. Constructed in-place structures may be of brick masonry or the concrete masonry type. Walls of structures should be constructed as designed and the dimensions of the structure should reasonably conform to that required. Masonry joints should be made with mortar completely covering the previously laid layer. The completed joint shall be finished neatly on the inside. If the structure is constructed of structural concrete, the requirements mentioned under the structural concrete section of this manual should be followed.

If precast or prefabricated units are used, all joints should be sealed to assure water tightness.

Careful attention should be paid to the backfilling operations to be sure no damage occurs to the structure and also to be sure compaction of the backfill material is obtained. Material should be placed and compacted in 6-inch level layers around the walls.

The grate or top for masonry or structural concrete structures should be adjusted carefully to the line and elevation required and supported on a bed of mortar. For prefabricated metal structures, the top should be placed and secured as required by the particular type being used.

Section 605 — Underdrain

Reference: AASHTO Guide Specifications Section 605, and applicable materials, Section 700.

This work consists of constructing underdrains using the type and size of pipe, or water-transporting filters, and granular material or filter fabric to the lines and grades shown on the plans or established by the Project Engineer.

The Inspector should know the specifications and special detail drawings for the type of underdrain to be constructed.

The locations of underdrains are usually determined by soils investigations prior to completion of the plans or during grading operations. Underdrains are placed to lower a high water table, to intercept and dispose of water seeping into the roadway from sources outside of the roadbed, or to intercept and control seepage from the backslope. Edge drains, placed parallel to and near the edge of pavement, are used to intercept seepage through the surface courses. Changes in design location or the selection of additional locations should be made by the Project Engineer and documented in writing.

Underdrain pipes should be placed with perforations down, except when their only purpose is to transport water. When their purpose is to carry water only, then a pipe without perforations should be used and thus it is unnecessary to place the granular material around the pipe. Blind drains are often installed, in which case the water table is lowered by the use of free-draining material.

Rigid inspection is required during construction of all types of underdrain. This assures that any slides from the sides of the trench are removed to ensure the filtering action of the granular backfill and that the holes in the underdrain pipe are not clogged with foreign material that would prevent the underdrain from functioning properly.

If equipment must cross underdrains after installation and prior to paving, the Inspector should insist on a rigid covering to protect the pipe from crushing and the granular material from contamination.

The Inspector should record the accepted quantity and location of all underdrain and should verify that all required tests and certificates of compliance are in the project record.

Section 606 — Guardrail

Reference: AASHTO Guide Specifications Section 606, and applicable materials in Section 700.

This work consists of the construction of guardrail of wire rope or cable, or cable-chain link, of beam, or of the rigid barrier type. The finished guardrail must be installed to the established lines and grades.

The Inspector should have full knowledge of the specifications and plans for the types of guardrail to be constructed. The Inspector should review the proposed locations of the guardrail as staked and ensure that the guardrail is properly placed to prevent the possibility of a vehicle running behind the guardrail into the hazard zone. All changes should be approved by the Project Engineer.

The pavement centerline should be established to align the guardrail posts. Holes for concrete posts are generally dug by auger, steel posts are generally driven but may be augered, and wood posts are either augered or driven. After the posts have been placed in auger-dug holes they should be backfilled and completed as specified. Posts should be set plumb, spaced as specified, and the top of the posts set to the design elevation. Be sure the rail laps are in the direction of vehicular travel and post-block assembly is correctly installed.

The Inspector should record the accepted lengths and locations of the guardrail sections and verify that the required tests and certificates of compliance are available in the project files at the time required.

Section 607 — Fences

Reference: Guide Specifications Section 607, and applicable materials in Section 700.

Fencing items consist of the furnishing and erection of woven wire, barbed wire, chain link fabric fences, and gates in conformity with the specifications and plans.

The Inspector should be familiar with all specifications and plans well in advance of any work by the Contractor and should arrange for all locations to be staked so the Contractor will not be delayed. All proposed locations should be reviewed and if changes either in location or type of fence are desirable, approval of the Project Engineer should be obtained. The Contractor should be furnished with a revised list.

Clearing and grubbing should be limited to the width necessary to erect the fencing and to provide a reasonable working room for equipment. Indiscriminate clearing should not be allowed.

The Inspector should check the installation or erection of all items of fencing to ensure that the posts are erected true to line, that the wire, fabric, and hardware are attached to the posts in the proper manner and at the proper elevation with the wire installed on a specified side, and make certain the posts are firmly installed.

The Inspector should record the accepted quantity for the types of fences and gates installed. Measurement for payment should be as stated in the specifications. The Inspector should verify that the certificates of compliance and necessary samples have been submitted by the Contractor.

Section 608 — Sidewalks

Reference: Guide Specifications Section 608, and applicable materials in Section 700.

This work consists of constructing bituminous concrete and portland cement concrete sidewalks.

The Inspector should understand the specifications for the type of sidewalk to be constructed. Location should be staked and checked well in advance of the work.

Usually sidewalks are constructed on a bed of coarse material or soil foundation which has been compacted to specification requirements.

All material specified to be tested must meet requirements before being incorporated into the work.

Bituminous sidewalks are normally constructed in one or two layers and compacted with a sidewalk roller where feasible or hand-tamped in places inaccessible to the roller. When constructing sidewalks adjacent to curbing, care should be taken to ensure the curb is not damaged or discolored. Wherever possible, the new sidewalk grade should meet existing driveway or walkway grades.

Portland cement concrete sidewalk forms should be perpendicular and strongly constructed and braced so that proper alignment and grade is maintained. Before portland cement concrete is placed, the bed course material should be thoroughly moistened so it will not absorb an excessive amount of moisture from the fresh concrete. Joint spacing, joint material, and reinforcing steel, if required, will be shown on the plans.

Proportioning of the portland cement concrete mix and the method of finishing and edging of the slabs are as outlined in the specifications. It is the duty of the Inspector to see that these requirements are carried out. Usually the Contractor has a choice of several methods of curing the concrete. Once the method of curing is selected, the requirements for the specific method must be adhered to. Bridges for pedestrians or vehicles may be required during the construction and curing periods for the protection of the sidewalk.

The Inspector should record the location and measurements for the quantity of sidewalks completed and accepted, and verify that all required tests and certificates of compliance are in the project files.

Section 609 — Curbs, Curb and Gutter, Paved Ditches, and Paved Flumes

Reference: AASHTO Guide Specifications Section 609, and applicable materials in Section 700.

This work consists of construction, setting, removing, and resetting of stone curbing constructed of quarried natural stone, cast-in-place concrete curbing, precast concrete curbing, dowelled concrete curbing, reflectorized concrete curbing, bituminous curbing, bituminous paved ditch, bituminous paved flumes, concrete paved ditch, and concrete paved flume.

The Inspector should have full knowledge of the specifications and design plans for all types of curbing, gutters, ditches, and flumes. The Inspector should review proposed locations of driveways, and, if changes are desirable, must obtain the approval of the Project Engineer and furnish the Contractor and Project Engineer with a revised location list. Well in advance of any work, locations should be staked and grades established.

The Inspector should inspect all precast and natural stone curb as it arrives on the project for conformity with the plans and specifications. The curb must be set on a firm foundation which has been compacted to density requirements. The base should be left a little above the bottom of the curb grade and then cut back to finish grade just ahead of the laying operation. This will assure that all base has its proper compaction. The curb should be protected after installation to protect alignment and grade. Any settlement or any misalignment shall be corrected before acceptance. Backfilling shall be conducted as specified.

Gutters, flumes, and paved ditches should be bedded in firm material, low enough so that water can flow over the sides without becoming trapped alongside the edge of concrete.

Portland cement concrete curb, gutter, flume, and ditch will require the same foundation as described above. The forms shall be strong and well braced. The concrete mix, joint material, and reinforcing steel, finishing, and curing must meet the requirements set forth in the specifications.

Bituminous concrete curb, ditch, and flume should be laid on the prepared surface with a reference line being used to produce the proper alignment. The finished curb should be checked for conformity with plans and specifications.

If slip-form placement is permitted by plans or specifications, all applicable requirements, as stated above, will apply.

The Inspector should record the accepted quantities of the various types of curb, gutter, ditch, and flume installed and should verify that the required tests and certificates of compliance, if any, are in the project files at the time required.

Section 610 — Turf Establishment

Reference: AASHTO Guide Specifications Section 610, and applicable materials in Section 700.

610.01 — Description

This item consists of the placement of topsoil when required by the contract, the soil preparation as necessary to receive the subsequent seed application, the application of lime and fertilizer, the application of seed, and the mulching of the seeded areas. Generally, the top layer of soil can be considered as a topsoil. It may be identified by a dark brown or gray color and contains a significant amount of organic matter. If the contract requires the salvaging and stockpiling of topsoil, the stockpile sites should be chosen at an early stage of the work and should be situated so as to facilitate the normal operational progress of the project and should also be chosen so as to cause minimal damage to vegetation proposed to be left undisturbed at the project site.

610.02 — General

Newly placed topsoil may slough off slopes when not bonded with the subsoil. Therefore, lateral scarification or marking of the subsoil which leaves ridges in the face of the slope may be required to prevent sloughing of slope areas.

Topsoil should not be handled when wet to the extent that it becomes densely compacted during placement. Finished surfaces of topsoiled areas should be in a condition to facilitate soil preparation. The depth of topsoil may be determined by digging test holes in back of the operation at frequent intervals.

Lime and fertilizer may be required by the plans, preparatory to seeding. The required rates of application should be closely followed.

Proper seedbed preparation is one of the most important factors in the development of the vegetative cover. The seedbed is the medium in which the seed germinates and from which the resulting vegetation, through its root system, secures moisture and nutrients. Soil should be broken up or loosened to a depth of five (5) inches to establish a proper seedbed with the top two (2) to three (3) inches receiving extra effort to pulverize it to a uniform consistency. Seedbed preparation is usually accomplished by the use of tillers and spring-tooth harrows to break up undisturbed soils and then further pulverization is obtained by the use of cultipackers, soil pulverizers, etc.

The application of seed, fertilizer, lime, and mulch may be placed by hydraulic means. Should the Contractor choose to make application in this manner, the specification requirements concerning this method should be followed carefully.

Should the Contractor choose to place seed, fertilizer, lime, and mulch by the dry method, the lime and fertilizer may be incorporated into the soil during the soil preparation; however, the application preparatory to soil preparation for both lime and fertilizer must be accomplished at the required rates and in separate placement.

Most seeding is accomplished by using mechanical seeders subject to the approval of the Project Engineer. The application of mulch shall be at the rate specified and should be securely anchored utilizing acceptable methods with use of the mulch tiller, asphalt emulsions, twine or netting, or other approved tie-down or adhesive materials. Completed seeding and mulched areas should be carefully observed periodically so as to make repairs or replace damaged seeded areas. In an effort to control erosion, completed slopes may be mulched during out-of-season periods with the actual seeding completed later during the specified seeding periods. Careful measurement should be taken of the items involved in the contract to provide adequate payment for the work accomplished.

When seeding dates are specified, no extension should be granted to these dates without the written approval of the Project Engineer.

610.03 — Records

Careful records should be kept on all items involved in turf establishment so that proper and adequate payment may be provided the Contractor.

Section 611 —Furnish and Plant Trees, Shrubs, Vines, and Groundcover

Reference: AASHTO Guide Specifications Section 611, and applicable materials in Section 700.

Planting plans are primarily a guide to the location for plants. The precise location of the plants on plans is not generally necessary for highway work, although accurate locations on the site are sometimes required. It is necessary in all cases to interpret the intent of the plans and meet local conditions such as having trees or other significant plants in the proper relation to utility lines or sight distances. The locations given on the plans should be scaled off, staked out on the ground; staking should be done by the Contractor or Department as indicated on the plans. In the event the Contractor provides the stakeout, adjustment may be made by the Project Engineer to properly meet site conditions.

If the Contractor is unable to furnish the specified kinds or sizes of plants, the permission to substitute should be approved by the Project Engineer. All data involving substitutions should be recorded in the project records.

All planting shall be accomplished within the season specified for the applicable types of materials to be planted. No planting shall be done in frozen ground, when snow covers the ground, or when soil conditions are unsatisfactory.

The Contractor must furnish the Project Engineer complete information concerning the source of supply for all plant materials. The Inspector should verify approval of the materials; however, approval of delivered nursery material to the project site shall not be considered as final acceptance.

It is the Contractor's responsibility to keep all plant material protected from damage and drying out at all times. Material delivered to the project site may be stored but must be protected at all times in strict conformity with the specifications. In an effort to temporily store delivered material at the project site, the Contractor may provide trenches in which the material may be heeled-in. The location of such pits or trenches should be determined before the plants arrive and should be excavated to sufficient size to accommodate the shipment. It is desirable to keep all of each kind and size of plant together with the last plant in the row carrying the identification tag. This plant should be the last plant removed from the pit.

Planting beds should be dug to the depth specified on the plans. As no greater depth of bed than shown on the plans should be provided and the plant roots must lie in a normal position, any roots of a length greater than the depth of prepared soil should be carefully removed before planting.

In the course of backfilling around set plants, the backfill soil must be a mixture of topsoil, loam or selected soil, and peat moss or peat humus, or as directed by the plans and specifications. Backfill should be placed around the roots of bare root plants in such a manner that no air voids are left and the roots lie in a natural position. The packing of soil under and against roots must be thorough.

When setting balled and burlapped plants after the backfilling operation has proceeded to a level to half the depth of the ball, the soil should be tamped and thoroughly watered. Any fibrous roots on the outside of the ball should be carefully placed in a natural position while placing the backfill. No ropes, burlap, or other material should under any circumstances be pulled from under the ball.

Fertilizing should be accomplished to the rate and type of fertilizer specified.

Water is an extremely important item in successful planting, and all plants should be watered during and immediately after planting, and at such intervals during the plant establishment period as directed by the Project Engineer. Watering during the establishment period, whenever indicated as a pay item in the contract, should be measured and paid for in accordance with the items in the contract.

Guying and staking is to be accomplished in accordance with the details shown on the plans as soon as the plant is in place.

Wrapping should be provided as required; however, wrapping should not be completed until after inspection of the plant by the Project Engineer. Subsequent to approval by the Project Engineer, the wrapping should be accomplished within the time limit established. The contract may allow the use of anti-desiccant spray at the direction of the Project Engineer. The anti-desiccant spray shall be applied within the time limits specified.

Pruning must be done before or immediately after planting in such a manner as to preserve the natural character of each plant. Pruning operations should be accomplished by thoroughly experienced personnel using properly conditioned equipment and in keeping with accepted horticultural practice. Cuts over 3/4-inch in diameter must be painted with an approved tree wound dressing.

Mulching must be furnished and placed in areas and to the depth specified. Mulch material may be wood chips, sawdust, or peat moss. Plants mulched with wood chips or sawdust should receive additional fertilizer as specified. The timing of the mulch placement is critical; therefore, the mulch should be placed within the time limit as specified.

During the planting operation, the Contractor should be required to use the utmost care to prevent unnecessary damage to disturbed areas and at the completion of the planting operation, the disturbed area is to be returned to its original condition by the Contractor.

During the establishment period, the Contractor should employ all possible means to preserve the plants in a healthy growing condition. This may include watering, cultivating, pruning, repair and adjustment of guys or stakes, and such other work as necessary or ordered by the Project Engineer. Dead or unsatisfactory plants should be promptly removed from the project. The length of the establishment period will be as indicated in the contract.

Careful records should be kept on all items involved in the furnishing and planting of trees, shrubs, vines, and ground covers so that proper and adequate payment may be provided the Contractor.

Section 612 — Mobilization

Reference: AASHTO Guide Specifications Section 612.

The timing of payment for this item as well as the schedule of payment should closely follow Section 612 of the Specifications.

Section 613 — Slope Protection

Reference: AASHTO Guide Specifications Section 613, and applicable materials in Section 700.

This work consists of the construction of slope protection using stone and filter blankets at locations required by the plans or the Project Engineer. Filter blankets can be granular material or filter fabric as specified.

The Inspector should know and understand the specifications for placement of stone slope protection and ensure that the slope has been brought to the required line and grade and the toe ditch prepaved prior to beginning placing operations. Placement of the slope protection and filter blanket will require care to prevent segregation of the aggregate.

The Inspector should record the accepted quantities of filter blanket and stone slope protection placed and should verify that the required test reports are in the project files.

Section 614 — Concrete Barrier

Reference: AASHTO Guide Specifications Section 614, and applicable materials in Section 700.

This work consists of constructing precast or cast-in-place concrete barriers.

The Inspector should have full knowledge of the specifications and design plans.

The alignment of forms for cast-in-place barrier must be smooth. The quality, shape, and alignment of slip-formed barrier must be equal to that of fixed-form barrier, otherwise, the slip-form machine should not be approved for use.

Precast barriers should be inspected upon arrival at the project for conformity with plans and specifications.

Concrete barriers must be placed on a firm foundation which has been compacted to density requirements and brought to required grade. Concrete surfaces which are to receive precast concrete barriers must be swept or flushed clean of sand and stone. Precast sections should be securely attached to each other per plan details.

The Inspector should record the accepted quantities of concrete barrier placed and should verify that the required tests and certificates of compliance, if any, are in the project files.

Section 615 — Erosion Checks

Reference: AASHTO Guide Specifications Section 615, and applicable materials in Section 700.

This work consists of the construction of gabions, grouted rock, straw bales, settling basins, and concrete checks in roadway ditches to prevent erosion.

The Inspector should have full knowledge of the specifications and plans for the types of erosion checks to be constructed. The Inspector should also be acutely aware of the need for timely installation of erosion checks to prevent siltation of streams or bodies of water.

The Inspector should ensure that all backfilling is performed as soon as practicable after construction. Straw bales should be well keyed into slopes and staked to prevent movement. Water should not be allowed to accumulate outside of checks.

The Inspector should record the accepted quantities of all erosion checks and should verify that all required tests and certificates of compliance, if any, are in the project files.

Section 616 — Riprap

Reference: AASHTO Guide Specifications Section 616, and applicable materials in Section 700.

Since channel changes usually after the hydraulic characteristics of a stream, riprap is frequently specified to protect property and roadways from erosion. When riprap is specified, the type and location will be specified on the plans. The plans show the general location but the definite limits must be determined based on field inspection.

Eddy currents often develop at the terminal ends of riprap. These terminals should be tied into some existing undisturbed features such as a boulder or rock outcrop. The alignment of riprap should conform as nearly as possible to the existing alignment of the stream. Any sharp changes in alignment causes changes in the characteristic flow of the stream and can increase the rate of erosion at some point along the stream or river.

The Inspector should record the accepted quantities of classes or types of riprap placed and verify that the required tests and certificates of compliance, if any, are in the project files.

Section 617 — Reference Markers

Reference: AASHTO Guide Specifications Section 617, and applicable materials in Section 700.

The Inspector should be aware of the need for early and accurate installation of reference markers.

Good public relations, as well as legal restrictions, require that trespassing by Contractor's personnel and equipment be avoided. Therefore, reference markers should be installed as early as practicable in the life of the contract so that lines may be readily established.

Care should be exercised to avoid any damage to reference markers and in the event damage occurs, the marker should be repaired or replaced immediately.

The Inspector should record the location type and quantities of reference markers and verify that the required tests and certificates of compliance, if any, are in the project files.

Section 618 — Traffic Control

Reference: AASHTO Guide Specifications Section 618, and applicable materials in accordance with the *Manual on Uniform Traffic Control Devices*, special provisions, and/or project Traffic Control Plan.

The control of traffic through construction projects is very important. Construction operations may inconvenience people who live along the right-of-way as well as those who must travel through a project built under traffic. Anything the Inspector can do to minimize this annoyance will improve public relations.

Section 619 — Erosion Mats and Bales

References: AASHTO Guide Specifications Section 619, and applicable materials in Section 700.

The erosion mat shall be of an acceptable type and placed on the designated areas immediately after the seeding or sodding operations have been completed. The Inspector should assure that the site is in an acceptable condition prior to placement of the erosion mat. The erosion mat shall be placed and anchored as specified.

The bales shall be placed at the contract designated areas to slow the water down and control erosion of ditches and other erodable areas. The Inspector should assure proper installation of the bales and the excavation of any necessary sumps is according to the contract. Once the slopes and ditches have stabilized and vegetative growth has developed sufficiently to control future erosion, the bales shall be removed and properly disposed of according to the contract.

When vegetative growth has not developed sufficiently to control future erosion and the contract work is completed and accepted, consideration should be given to leave the bales in place until such time that the vegetative growth can control future erosion. Then, the bales could be removed and properly disposed of by the contractor.

Sod shall be of an acceptable type and condition. The Inspector shall assure that the sod is placed within the specified time frame on an acceptable site. The sod should be anchored and watered according to the contract. The inspector should assure that the sod is receiving an adequate amount of water to prevent damage to the sod for the duration specified.

The inspector should assure that the appropriate seed and fertilizer is applied at the specified rate on the designated areas.

Section 620 — Filter Fabric

References: AASHTO Guide Specifications Section 620, and applicable materials in Section 700.

The filter fabric shall be of the type specified. The Inspector should assure the appropriate type of filter fabric is being placed according to the contract at the designated locations.

The Inspector should be informed as to the Contractor's personnel responsible for traffic control surveillance so that necessary action may be taken with the least delay.

The Inspector should evaluate constantly the Traffic Control Devices (TCD) in light of the Contractor's operations. Modifications may be needed to ensure safety. Knowing when TCDs are not required is just as important as knowing when they are required. The effectiveness of all TCDs is reduced if one is left in place after the need for it passes. This leads drivers to believe they can ignore other TCDs in construction zones. When TCDs are no longer required, they should be removed.

The Contractor should maintain all TCDs in a satisfactory condition. Cleaning of TCDs must be done periodically to improve their visibility at night. A night inspection of TCDs is necessary to evaluate them as this is the period when the traveling public must rely more heavily on the TCD.

Unnecessary pavement markings should be eradicated as soon as possible. Traffic movement is influenced to a great extent by pavement markings; therefore, misleading markings can be a significant contributing cause of traffic accidents. Additionally, temporary pavement markings should be placed where the travel path has been changed and where traffic is to operate on new pavement overlays.

The Project Engineer, or designated representative, should review the condition of the project at least twice daily, once in the morning and again before leaving in the evening, to ensure that barricades, warning lights, signs, barrels, and other traffic control devices are in the proper location and operating correctly. Any unusual situation which may require changing, adding, modifying, or deleting TCDs should be brought to the attention of the Contractor. Any irregularities in the operation of the detour items or defects in the detour surface should also be brought to the Contractor's attention for prompt correction.

The public's impression of a project is greatly influenced by its flagging operations. If the flaggers are neat, courteous, and efficient, the public will accept inconvenience of construction with little question. A rude, discourteous, or lazy flag person will create a potentially dangerous condition, especially if the public questions their credibility. One of the primary rules for all flaggers is to dress with official clothing and accessories. Above all, a flagger must be familiar with and utilize proper flagging procedures.

Proper positioning of a flagger is important to the safety of the public, the Contractor's work force, and personal safety of the flagger.

DIVISION 700 — MATERIALS CONTROL

Section 701 — Purposes of Materials Control

701.01 — Compliance with Specifications

An adequate and effective system for control of materials used in a project is absolutely essential to ensure that the materials furnished and the completed work produced by the Contractor conform to the contract requirements.

The system to determine compliance may vary in form. The system may be designed with full control and all sampling and testing being performed by the Department or be designed to combine a Department conducted program for acceptance sampling and testing with demonstrated and approved quality control procedures conducted by others.

In either case the acceptance of materials incorporated in the work must be based on process inspection, material use, sampling, testing and measurements by the Department in order to provide complete and positive verification of compliance. Without this verification of compliance the Department cannot fulfill its responsibilities under the contract for the commitment of public funds.

701.02 — Uniform Relations with Contractors

The specifications and plans provide a basis for bidding by contractors since they define the minimum requirements that are to be met. The Contractor is committed to furnishing materials and completed work that will equal or exceed such requirements.

The Project Engineer must assure, through materials control measures, that the Department is receiving what it is entitled to under the contract. To accept anything less would not only be a disservice to the Department but would be giving undue advantage to the Contractor. Other contractors who had bid on the same work could contend that they would have offered a lower bid had they been able to anticipate that materials or work outside the contract would be accepted.

In exercising materials control, it is essential that it be applied uniformly by all Engineers and Inspectors from project to project so that all contractors and suppliers are treated alike. Difference of interpretation of the contract, both on the part of the Contractor and on the part of the Department personnel regarding minimum acceptance values must not be allowed to happen. This construction manual provides guidance to the Project Engineer and Inspector on the interpretation of the contract and the application of the materials control measures.

701.03 — Documentation for Expenditure of Public Funds

When payment is to be made to the Contractor for materials furnished and work performed, the duly designated Department officials will authorize disbursement of public funds for this purpose. The officials authorizing payment must depend upon others for evidence to support the expenditure. Through the materials control process, the Project Engineer will acquire substantiating data in the form of tests, inspection records, and measurements to justify acceptance of the Contractor's work. Thus, the Project Engineer can assure and furnish documentation to the officials responsible for authorizing payment that the Contractor has fulfilled contract obligations and is entitled to full payment.

In case of failure to meet the material quality requirements, the materials control data will constitute the basis for rejection of work deemed unfit for acceptance, or it may be the basis for its acceptance upon appropriate contract price adjustment where this is permitted under the provisions of the contract.

Complete records including test and inspection reports covering acceptance or rejection should be maintained in the project files, while the necessary copies should be furnished to the appropriate headquarters personnel as needed for verification and as supporting evidence for payment documents. A standardized set of instructions and examples in the form of a "Documentation Manual" can be of great value to outline documentation procedures for field personnel. Department field personnel can use this convenient reference as a guide for documenting any of the various items of work under their supervision.

Section 702 — General Procedures for Materials Control

702.01 — Specifications and Their Application

Specifications are prepared to define minimum requirements for materials that are considered suitable for specific purposes. When materials are furnished to meet these specifications, tests must be made on the material and compared against the requirements to determine acceptability.

Sampling, testing, and taking of measurements must be done strictly in accordance with the procedures that are prescribed in the specifications and referenced material sampling and testing procedures. The points at which the specifications, the sampling, testing, or measurements are to apply must also be carefully observed. Failure to comply with any of the required procedures can result in serious discrepancies in results.

702.02 — Inspection — General Procedures

A. Inspector's Responsibility. The inspector on the project or at a plant plays a vital part in the control of materials. It is the responsibility of the inspector to see that no material goes into the work that has not been inspected and tested or approved by Department inspection personnel. Where the materials delivered to the project have been manufactured by a detailed closely controlled process approved by the Department, their incorporation in the work may be allowed without continuous Department control sampling and testing.

When the material delivered to the project for incorporation in the work has been sampled and tested by Department inspectors during interim steps in the manufacturing process, the project inspector must assure that the material test results meet specification requirements before the material is placed.

Acceptance sampling and testing by the Department must still be performed, however, before payment can be released for material provided under either situation.

The Department inspectors must know which materials must be sampled, when and where samples must be taken, the size of samples required, the proper methods for obtaining samples, and methods of field testing.

B. Manufactured Product Inspection and Acceptance. Incorporation in the work of some materials may be permitted provided the Department allows the use of certificates of compliance from the source manufacturer accompanied by test results showing specification compliance. The certificate of compliance must state the contract requirements for the product material and that the requirements have been met. The certificate must be signed by a legally responsible member of the manufacturing firm and identify the quantity of material by lot number for which the certification is applicable.

The nature of the material in this case is such that the cost of the testing equipment and nature of the testing procedures are too prohibitive to be accomplished at the project level or Department laboratory satellite facility and the industry quality control procedures used show consistent satisfaction of the specifications. Examples of materials in this category would be: metal products, Portland cement, asphalt, chemical products, electrical products, prestressed structural members, and others which involve the same level of manufacturing accuracy control. Examples of material that would not fall in this category are: asphalt cement mixtures, Portland cement or Portland pozzolanic concretes, aggregates and soils.

Source accepted products delivered to the project must still be sampled, possibly on a reduced frequency identified by the Department Materials Engineer, and forwarded to the Department testing laboratory for testing to verify that the specification requirements of the product have been met and provide the basis of Agency acceptance.

The source acceptance of products must be approved by the Department Materials Engineer. The preparation and updating of a list of approved product manufacturers and the accepted products will be provided by the Department Materials Engineer. Continuous updating of the list of source approved products is necessary based on the results of verification testing of the product qualities required by the specifications. Quarterly publishing of the approved list and distribution to the Department field offices is suggested. Immediate communication both ways, between the Department laboratory and projects, is needed in the event of noted failure of the material to meet specification requirements.

Approval of materials at the source for delivery to the project must be properly tagged or marked on the material or its container to provide positive identification when shipped to the project. Prior to material incorporation in the work, the project inspector must verify that a complete materials test report indicating compliance with the specifications and certificate of compliance has been provided to the Department for the material delivered to the project. A visual inspection and dimension check, where applicable, is also to be made by the project inspector to assure that the material has not been damaged during shipment to the project and is dimensionally correct for the required use.

Full payment for the material use under the appropriate specifications should not be provided until approved test results or certification is obtained.

Inspection by Department inspectors of source acceptance materials fabrication is necessary where those materials must undergo further construction modification before acceptance of the in-place material can be provided by the Department. Examples of materials in this category are: metal structural members, reinforced and non-reinforced precast concrete products, welds, coatings, and pre-stressed materials. The results of this inspection and copies of applicable manufacturers test are to be forwarded, with the properly identified material to the project.

The inspector at the plant or project will need to take samples of the material, where feasible, to serve as verification of the manufacturer's process control and acceptance testing by the Department.

Copies of the project material test results, material certification and release for payments are to be placed in the project records by the inspector.

C. Inspection of Manipulated and Processed Materials. When materials are combined into mixtures at plants or otherwise processed at or near the project site, the ingredients are to be properly sampled and tested by the Department at the plant or process point and the mixture or the processed item further sampled and tested by the Department for acceptance payment as close as practical to final incorporation in the work. When tests on samples cannot be performed in the field, the samples are to be packaged and transported, as directed by the Materials Engineer, to the Department laboratory equipped to perform the test.

If the contractor or producer has a Department approved and operational quality control plan and has demonstrated consistent uniform acceptable materials production or mixtures, the frequency of acceptance sampling and testing may be reduced at the production plant or process point by the inspector. The Materials Engineer will identify the modified sampling and testing procedures.

The Department inspector at the plant is to assure that provisions of the approved quality control plan for the contractor or producer are followed, including sampling and testing procedures identified in the plan, documentation of test results, and plant condition and operation by periodic inspection throughout the day.

The specification requirements of the processed material or mixtures tested by the Department at the plant must still be met. The test results showing failing material are used as the basis for rejection of the material or mixture as with test results from the point of incorporation in the work. The contractor or producer test results are not to be used as the basis of acceptance for payment by the Department. Sampling and testing of materials for acceptance and payment must be performed by the Department to assure the specification quality requirements have been met.

Periodic Department review of the contractor or producer quality control plan effectiveness to provide a consistent uniformly acceptable product will be necessary. This review period, established by the Materials Engineer, for which the plant is covered must be provided to the pertinent Department inspectors since the inspector must know whether the reduced sampling and testing frequency referred to above, is applicable for the plant. The maximum time between Department reviews of the quality control plan effectiveness is one year.

Communication between field and plant inspectors must be established to coordinate material quality control to meet the specification requirements. Test result data from both inspection points must be freely exchanged.

D. Project Inspection. The Project Inspector must identify and check all materials received on the project before they are incorporated into the work, must perform tests and inspection required, and assure that Department acceptance tests and inspection reports show that the material is in compliance with the contract requirements.

Materials that have been inspected prior to delivery must be reexamined for any damage or contamination that may have occurred during or after delivery or for any defects that may not have been observed in the original inspection. Defects or contamination may be cause for rejection in spite of prior inspection, unless they can be satisfactorily remedied.

All materials received on the project without prior inspection and approval are to be inspected, sampled, and tested by the Project Inspector. If the required tests cannot be made at the project, the Inspector will send the appropriate samples to the Central Laboratory for testing, and upon notification of test results the Inspector is to either accept the material for payment or use, or reject it. The Contractor shall be promptly notified of rejection of unacceptable material.

Materials such as aggregates that are produced at or near the job site and are subject to inspection as produced will be sampled and tested for acceptance and payment by the inspector on the project or at the production plant. Materials such as asphalt mixes, Portland cement concrete or Portland pozzolanic concrete, and stabilized aggregates or soils that need to be sampled and tested at both the plant and at the project to assure contract compliance of the mixes, will need to have the sampling and testing coordinated to assure material locations identification is known and to correlate testing results. Tests necessary to measure contract material quality requirements that cannot be performed at the project level will be sampled by the inspector and sent to the appropriate Department laboratory identified and in the manner directed by the Materials Engineer.

Along with examining and checking all materials brought onto the job, the Inspector should maintain a continuing visual inspection of the Contractor's operations where the materials are handled and incorporated into the work. Any procedures that result in damage or change in any material to the extent that it will fall outside the specification limits should not be permitted to continue, and the materials so affected should be rejected.

When statistical quality control specifications are used they will include consideration of material quality variation in the establishment of acceptable test result values. This type specification identifies allowable ranges in test values considering the probability of occurrence of failing test results. Material in place which has been mishandled or damaged by the contractor, or which is obviously unacceptable is not to be accepted and is to be corrected or removed in order to provide improved specification compliance of the end product.

Diligent inspection of the work in progress and of each successively completed portion is important if there is to be real assurance that when the work is finished all parts of it will be acceptable. Sampling and testing without observations at the same time reduces this degree of quality assurance.

E. Miscellaneous Materials and Small Project Material Quantities. When small quantities of certain materials are included in the project contract and the cost of sampling and testing would be excessively high compared to the project and item cost, certification of compliance with the specifications by the supplier or producer may be used as the basis of material acceptance. As stated previously, inspection of the material and placement must still be provided and obviously unacceptable material, delivered or in place, rejected.

This acceptance procedure is not to be used where the functional requirements of the material are critical, such as bridge deck or load bearing Portland cement concrete use and safety devices.

Use of the certification must be based on Department wide published criteria established by the Materials Engineer identifying the material for which certification will apply and identifying the project quantity limits for each material. Updating of the list of materials and quantity limits after periodic evaluation of material performance by the Materials Engineer, will be necessary.

As an aid to evaluate performance of certified material compliance items, occasional sampling and testing for spot checking is suggested.

Examples of material small quantities in such cases would be as follows:

Hot mix asphaltic concrete

- Wearing course — 500 tons
- Binder and base course — 1000 tons

Aggregate Base — 600 SY — 6″ depth

Portland cement concrete paving — 600 SY

Structural concrete — 50 CY
Miscellaneous Portland cement concrete
Non-traffic area use — 150 CY
Reinforcing steel bars or mats — 1000 lbs.
Liquid asphalt — 100 tons
Earthwork — 5000 CY

F. Disposition of Failing Material. In the case of a sample test result not meeting, or not in conformity with the specification; the material represented by the sample shall be rejected or remedied, or otherwise handled in accordance with the specifications. Failing material that has not been finally incorporated into the work and can be remedied by further processing to bring it within requirements may be accepted after having been corrected.

If completed work is found to contain material that is not in conformity with the specifications, and the specifications make proper provision for doing so, the Project Engineer will make a determination as to whether the completed work will perform satisfactorily and can be accepted on the basis of a suitable adjustment of the contract price, or whether it is so defective as to require removal and replacement or repair if feasible. The action taken in such a situation must be fully documented by reports and records covering samples, tests, measurements, corrective action, if taken, any supplementary agreement or price adjustment, and justification for acceptance or rejection.

G. Records. As previously stated, the Project Engineer must provide adequate documentary evidence of specification compliance to support acceptance of and payment to the Contractor for all materials furnished and work completed. Therefore, the Inspector on the project must prepare the appropriate reports covering all tests, measurements, and inspections performed in the field. Reports must also be available covering items inspected and tested at source, as well as test performed in the Central Laboratory, and certifications of compliance.

All these reports are to be made a part of the permanent project record and copies are to be furnished to the appropriate district or headquarters personnel for review and appraisal in connection with final project certification. The Inspector should also record on the inspection reports, or project diary, any pertinent observations regarding the use of materials or the Contractor's operations that are worthy of special mention.

Section 703 — Sampling

A. Representative Samples. To obtain reliable, representative samples is one of the most critical and exacting tasks the Inspector has to perform. If a sample is to be worth anything, it must be truly representative of the particular unit or quantity of material that is to be evaluated.

In sampling aggregates, for example, it is often difficult to secure good samples, largely because of the tendency of particles of different sizes to segregate. This occurs when any material composed of a mixture of particles that vary widely in size is deposited in a pile such as a conical stockpile. The coarser particles will roll to the outside and accumulate around the outside and bottom of the pile. It is therefore very difficult to get a single sample that is typical of the gradation of the entire pile.

Similarly in liquids, such as asphalt materials in large tanks, there may at times be stratification, producing appreciable variation in material at different levels in the tank. Consequently, a sample taken at a single level may not be really representative of the entire volume.

Therefore, when sampling any material or finished work, the Inspector must be alert to the variation that may exist and take every precaution to ensure that the samples taken are as representative as possible by following proper specified sampling procedures.

B. Number of Samples. When large volumes or areas are to be sampled, it is necessary to divide the material into portions or lots, each of which may be sampled separately. Results of tests on the different samples will show not only which lots meet the requirements, but also how uniform or variable the material is from lot to lot or from place to place, or time to time.

The number of samples to be taken will depend on the nature of the material, its inherent uniformity or variability, and the criticality of the material in the location proposed for use. For example, granular material used for backfilling of swamp excavation would not be as critical in gradation as similar material used for subbase under a flexible pavement. The frequency of sampling and testing for control could therefore be considerably less in the former case than in the latter. Also, if the tests on a series of samples show that a given material is very uniform and well within requirements, the number of samples taken could be less than for any material that is variable or on the borderline of failure. Uniformity, however, should not be cause to stop sampling and testing for compliance with the specifications.

The rate of sampling should in any case be sufficient to show clearly in the records that it conforms with specification requirements and that it is in keeping with the importance of the item.

In general, there are three different ways which may govern the rate of sampling:

1. sampling regularly at some predetermined interval of time, distance, or volume;
2. selecting points or items to sample that are obviously or more likely to be deficient or suspect;
3. sampling by pure change, or random selection of the time or place of sampling.

None of these alone is necessarily the best procedure for all situations, but generally the sampling should be based on a combination of at least two.

A combination of at least two of the above procedures is recommended to provide the Agency, as facility owner and steward, full confidence that the materials used are of a quality within normal bounds of material variability and that there is no material used, as best can be determined, that is outside the normal bounds of variability and the specification limits. It should be borne in mind that the purpose of sampling and testing of material is to establish that the materials to be used meet minimum standards and is not a full measurement of the quality of the completed facility since construction processes will have a significant influence on the product also.

Taking samples only at a predetermined constant interval, as in item one above, does not take into account the many factors that have some bearing upon when and where to sample. However, experience has demonstrated rather well that maintaining minimum rates for sampling of different kinds of materials and types of work is desirable to provide good control and help assure a good finished product.

Under the procedures of item two above, the Inspector on the job would have no guidance except to exercise judgment and experience through continual visual inspection and observation of the Contractor's operations. Thoughtful inspection and selection of points to sample or test will do much to detect deficiencies, and the results of such tests can be the basis for improvement of procedures that may prevent such deficiencies from recurring. But along with this, there should be some minimum frequency of sampling and testing maintained.

Selection of samples on the basis of chance, or at random, as in procedure three above, is the method by which every part for a lot or unit has an equal chance to be selected. This is the preferred method for providing a statistical approach to sampling. When the general level of quality of work and materials is high and reasonably uniform, the random procedure will provide representative sampling. Selection of the size of units or lots for the random sampling to conform with the minimum rates will provide the adequacy desired. However, where there are abrupt or extreme variations that may be critical and may endanger the completed work, the Inspector should be especially diligent in visual inspection of the Contractor's operations. Selective sampling of any obviously questionable items should be done to supplement the random sampling program.

Schedules of suggested sampling frequencies for major work items and materials are included in this section and consists of the following:

1. Process Control

This is a schedule of suggested sampling frequency for process control purposes, whether done by the Department or done by the material producer under an approved quality control plan to serve as a guide for the inspector. The suggested frequencies shown are base figures that should be used by agencies that do not have historic data to demonstrate that a reduced or increased frequency of sampling and testing provides quality assurance for the materials and producers normally used. The use of the reduced frequency sampling and testing would be based on historic data of normal producers used by the Department, and good materials. If, however, the materials or operations are variable and good control is difficult to maintain, samples should be taken at greater frequency until the desired uniformity is obtained and can be maintained.

2. Acceptance

A schedule of suggested sampling frequency for Department acceptance of material for payment and has been developed and included in this section to serve as a guide for the inspector. The suggested frequencies shown may be modified upward or downward based on historic data of Department conducted acceptance test results reflecting the quality of the material normally used and quality of contractors working with the material.

3. Independent Assurance

A schedule of suggested sampling frequency for Department independent assurance sampling and testing review has been developed and included in this section to serve as a guide to evaluate the quality of the sampling and testing procedures and equipment. The suggested frequency may be adjusted upward or downward for any material, however, the lowest frequency of independent assurance testing is not to be less than 10 percent of the acceptance tests, or once per month per item of acceptance testing.

Section 704 — Sampling Procedures

Samples should always be taken in accordance with the procedure that is prescribed, so there will be no question about it being valid from that standpoint. This can become very important in case of a test failure which may be the basis for rejection of an item. If the sample is questionable as to the manner in which it was taken, there is legitimate cause for contention on the part of the Department or Contractor that the result is not reliable or acceptable as a basis for rejection or acceptance.

Standard methods of sampling have been developed by AASHTO for many materials, and these are commonly specified and used when applicable. Where AASHTO methods are not available, other standards that have been developed such as ASTM are frequently specified. In cases where neither of such standards is available and specified, the procedure to be used should be one that has been adopted by the Department and is so identified that the Contractor and materials suppliers will know precisely what is expected of their material. The specific methods of sampling are usually referred to by reference numbers in each specification. Many of the test procedures allow sampling to be done in several locations. The sampling location to be used by the Department inspector will be as established by the Department Materials Engineer.

The sampling of aggregates of various kinds and other granular materials such as cement, mineral filler, and similar items, comprises much of the sampling operation performed by the Inspector on many jobs. As previously noted, segregation may have a great influence upon the gradation as determined by tests. Therefore, it is especially important to use extreme care in obtaining the samples of these materials.

Where a large quantity of material is involved, such as in a stockpile, a number of separate samples should be taken from different areas to represent the variations that may exist.

The size of sample needed for the various tests on a given material is usually given in the standard procedures for the test. In the case of aggregates, the sample size is dependent largely upon the maximum and predominant particle sizes. The larger the maximum size, the larger the sample should be. The recommended weights of samples are included in the *AASHTO Methods of Sampling and Testing*.

Where a large composite sample is made up of a number of smaller samples, it may be necessary, for the purpose of testing, to reduce the sample to the quantity desired for the tests. This should be done either by the quartering method or by use of a sample spitter.

For other materials, the size of the total sample must be large enough to provide the quantity needed for all of the specified tests.

Section 705 — Testing Procedures

As with sampling, standard procedures for testing have been developed by AASHTO and ASTM which are commonly specified for use where applicable. When standard methods are not available or may not be considered suitable, the Department may have adopted some modifications or methods of its own development.

Therefore, testing, as well as sampling, should be done strictly in accordance with the specified procedure or as otherwise agreed upon as the basis for evaluating the acceptability of a material. In certain materials specifications, some of the test methods that are specified may be designed for use in a Central Laboratory and may not be adaptable, as written, to use for field testing.

Field test methods may sometimes be modifications of standard procedures. When that is the case, rejecting of material on the strength of a field test failure should be done only if the field test method is specified or if there is good documented correlation between the field test method and the specified laboratory method. When a standard method is specified, this is the referee procedure which must govern in the event of any controversy regarding rejection.

The Inspector must always remember that the contract is a "two-way street." What is stated in the specifications is a commitment on the part of the Department, as well as on the part of the Contractor, to perform its obligations as stipulated. The Contractor has the right to expect conformance by the Department as much as the Department expects it from the Contractor.

Section 706 — Independent Assurance Sampling and Testing

The intent of this program is to serve as an unbiased and independent verification of the procedures used and of the test results obtained by project personnel in their regular inspection and testing activities. The results of independent assurance tests are not to be used as a basis of material acceptance. The program is a check on sampling and testing procedures and equipment and is designed to supplement the regular acceptance sampling and testing program. It involves a separate and distinct schedule of sampling, testing, and observations.

The personnel who perform these independent sampling, testing, and observation duties are to be qualified and well-trained and shall represent the Department Headquarters or Laboratory and shall have not direct involvement in project control or management. However, cooperation of the project personnel is required. Also, the testing equipment used must not be the same equipment used for acceptance testing.

Observation of Department conducted acceptance testing by personnel responsible for independent assurance may be allowable for not more than 20 percent of the independent assurance samples and tests.

An important part of this program is the careful documentation of all activities, prompt evaluation of test results by comparison of such test results with other available test data, the documentation and explanation of all deficiencies and anomalies and possible corrective actions taken.

Comparison of the acceptance test results with the independent assurance results, evaluation of any differences, and the determination of sampling and testing corrective actions needed is to be done by a responsible person designated by the Department Materials Engineer.

Suggested allowable deviation values between acceptance and independent assurance test results prior to corrective action or explanation are as follows:

Materials	*Test/Variation*						
Aggregates for	Gradation Sieve Size						
Portland cement	1″&3/4″	1/2″&3/8″	#4	#8&#16	#30	#50&#100	#200
concrete	8%	8%	4%	3%	3%	3%	1.5%
Aggregates for	Gradation Sieve Size						
Treated and	1″&3/4″	1/2″&3/8″	#4	#8&#16	#30	#50&#100	#200
Untreated Bases and Subbases	6%	6%	5%	4%	3%	3%	1.5%
Aggregates for	Gradation Sieve Size						
Hot Mix	1/2″&3/8″	#4	#8&#16	#30	#50&#100	#200	
Asphalt plant production	4%	4%	3%	3%	2%	1.5%	
Soil and	Moisture Content						
Aggregates	0.5%						

Materials	*Test/Variation*			
Portland Cement concrete	Air Content 0.5%	Slump 1/2″	Wt./Ft. 1lb.	
Earthwork and Base/Subbase Compaction	Density In-Place 3 lb.	Moisture Content In-Place 1%	Proctor Density One-Point 2 lb.	
Hot Mix Asphalt Concrete	Asphalt Extraction 0.3%	Maximum Theoretical Density 1/2 lb.	Density (Nuclear) 3 lb.	Density (Cones) 1/2 lb.

Section 707 — Suggested Sample and Test Frequencies

As stated earlier in this section, the inspector's responsibilities include knowing which materials must be sampled and tested before being incorporated in the work or accepted for payment purposes. The inspector must also know where, when and how to take samples, what size samples, what field tests are required and how the tests are to be conducted.

The following table of material testing and sampling frequency includes only the basic construction materials normally used in highway construction. The tests and frequency shown are intended as a guide for use by highway Departments in preparing their own individual materials sample and test procedures. Each individual Department may wish to modify the table to meet their respective needs or requirements.

TABLE OF MATERIALS TESTING AND FREQUENCY

	Without Contractor Process Control		With Contractor Process Control			
Earthwork	**Acceptance Testing**	**Independent Assurance Testing**	**Process Control Testing**	**Verification/ Check Testing**	**Acceptance Testing ***	**Independent Assurance Testing**
Soil Identification	1/excavation strata		1/excavation strata	1/excavation strata		
Optimum Moisture—Maximum Density	1/soil class/project	1/project	As needed to control operation	1/lanemile	1/soil class/ project	1/project
Compaction	1/1000 ft./lift	1/2 weeks	As needed to control operation	1/2 weeks	1/2000 ft./ lift	1/month

*Acceptance testing is to be performed at the process control frequency until uniform material is produced.

TABLE OF MATERIALS TESTING AND FREQUENCY

	Without Contractor Process Control		With Contractor Process Control			
Bases/Subbases	**Acceptance Testing**	**Independent Assurance Testing**	**Process Control Testing**	**Verification/ Check Testing**	**Acceptance Testing ***	**Independent Assurance Testing**
Aggregate & Stabilized Aggregate						
Gradation	1/1000 tons	1/2 week	1/1000 tons	1/2 week	1/2000 tons	1/month
Moisture/ Density	1/1000 ft.	1/2 week	1/1000 tons	1/2 week	1/2000 tons	1/month
Stabilizer Content	1/1000 tons mix	1/2 week	1/1000 tons mix	1/2 week	1/2000 tons	1/month

*Acceptance testing is to be performed at the process control frequency until uniform material is produced.

TABLE OF MATERIALS TESTING AND FREQUENCY

	Without Contractor Process Control		With Contractor Process Control			
Bases/Subbases	**Acceptance Testing**	**Independent Assurance Testing**	**Process Control Testing**	**Verification/ Check Testing**	**Acceptance Testing ***	**Independent Assurance Testing**
Soil & Stabilized Soil						
Soil Classification	1/source		1/source	1/source		
Moisture/ Density	1/1000 LF	1/2 weeks	1/1000 LF	1/2 weeks	1/2000 LF	1/month
Cement Content	1/10 ton cement	1/2 weeks	1/10 ton cement	1/2 weeks	1/100 ton cement	1/month
Lime Content	1/10 ton lime	1/2 weeks	1/10 ton lime	1/2 weeks	1/100 ton lime	1/month
Asphalt Content	1/20 ton AC	1/2 weeks	1/20 ton AC	1/2 weeks	1/200 ton AC	1/month
Aggregate Content	1/1000 ton aggregate	1/2 weeks	1/1000 ton aggregate	1/2 weeks	1/2000 ton aggregate	1/month
Compressive Strength (Lime/ Cement Stab.) 10 day age	1/1000 SY		1/1000 SY	1/2500 SY	1/5000 SY	

*Acceptance testing is to be performed at the process control frequency until uniform material is produced.

TABLE OF MATERIALS TESTING AND FREQUENCY

	Without Contractor Process Control		With Contractor Process Control			
Aggregate Production **	**Acceptance Testing**	**Independent Assurance Testing**	**Process Control Testing**	**Verification/ Check Testing**	**Acceptance Testing ***	**Independent Assurance Testing**
Coarse Aggregate						
Gradation	1/1000 ton	1/2 weeks	1/1000 ton	1/2 weeks	1/2000 ton	1/month
Soundness	1/6 month				1/6 month	
Abrasion	1/6 month				1/6 month	
Deleterious Substance	1/6 month				1/6 month	
Fine Aggregate						
Gradation	1/500 ton	1/2 weeks	1/500 ton	1/2 weeks	1/5000 ton	1/month
Particle Size Analysis	1/6 month				1/6 month	
Sand Equivalent	1/6 month				1/6 month	
Blend Aggregates						
Gradation	1/1000 ton	1/2 weeks	1/1000 ton	1/2 weeks	1/2000 ton	1/month

*Acceptance testing is to be performed at the process control frequency until uniform material is produced.
**Based on mined material. Quarry operation test frequency for process control and acceptance could be reduced to account for historic quality control results.

TABLE OF MATERIALS TESTING AND FREQUENCY

	Without Contractor Process Control		With Contractor Process Control			
Asphalt Paving	**Acceptance Testing**	**Independent Assurance Testing**	**Process Control Testing**	**Verification/ Check Testing**	**Acceptance Testing ***	**Independent Assurance Testing**
Asphalt Content (Extraction)	1/500 tons	1/2 weeks	1/500 ton	1/2 weeks	1/day (1/2000 ton)	1/month
Mix Aggregate Gradation (Wash)	1/500 tons	1/2 weeks	1/500 ton	1/2 weeks	1/day (1/2000 ton)	1/month
Aggregate Sp. Gravity	1/month		As needed to control operation	1/month		
Voids in Mineral Aggregate	1/day		1/day	1/2 weeks	1/2 weeks	
Maximum Theoretical Mix Density			As needed to control operation	1/2 weeks		
Aggregate (Fracture Faces)	1/month		As needed to control operation	1/month		

*Acceptance testing is to be performed at the process control frequency until uniform material is produced.

TABLE OF MATERIALS TESTING AND FREQUENCY

	Without Contractor Process Control		With Contractor Process Control			
Asphalt Paving	**Acceptance Testing**	**Independent Assurance Testing**	**Process Control Testing**	**Verification/ Check Testing**	**Acceptance Testing ***	**Independent Assurance Testing**
Compaction	5 locations/day	1/2 weeks	As needed to control operation	1/2 weeks	5 locations day	1/month
Anti-Strip	1/day		1/day	1/2 weeks	1/month	
Marshall Mix Design						
Voids				1/day		
Stability				1/job mix formula or 10,000 ton		
Flow				1/job mix formula or 10,000 ton		
Hveem Mix Design						
Voids				1/day		

*Acceptance testing is to be performed at the process control frequency until uniform material is produced.

TABLE OF MATERIALS TESTING AND FREQUENCY

	Without Contractor Process Control		Supplier Process Control			
Asphalt Materials	**Acceptance Testing**	**Independent Assurance Testing**	**Process Control Testing**	**Verification/ Check Testing**	**Acceptance Testing** *	**Independent Assurance Testing**
Asphalt Cement Quality OR Cut Back Asphalt Quality OR Emulsified Asphalt Quality			1 per lot	Agency wide review—1/supplier per month		

*Acceptance testing is to be performed at the process control frequency until uniform material is produced.

TABLE OF MATERIALS TESTING AND FREQUENCY

	Without Contractor Process Control		With Contractor Process Control			
Asphalt Paving	**Acceptance Testing**	**Independent Assurance Testing**	**Process Control Testing**	**Verification/ Check Testing**	**Acceptance Testing**	**Independent Assurance Testing**
Stability				1/job mix formula or 10,000 ton		
Swell				1/job mix formula or 10,000 ton		
Aggregate Surface Area & Absorption				1/job mix		

TABLE OF MATERIALS TESTING AND FREQUENCY

	Without Contractor Process Control		With Contractor Process Control			
Portland Cement Concrete Pavement	**Acceptance Testing**	**Independent Assurance Testing**	**Process Control Testing**	**Verification/ Check Testing**	**Acceptance Testing ***	**Independent Assurance Testing**
Mix Design Gradation						
Coarse Aggregate	1/500 ton aggregate	1/2 weeks	1/500 ton aggregate	1/2 weeks	1/2000 ton aggregate	1/month
Fine Aggregate	1/500 ton aggregate	1/2 weeks	1/500 ton aggregate	1/2 weeks	1/2000 ton aggregate	1/month
Moisture Content**	Daily/stockpile aggregate	1/2 weeks	Daily/ stockpile aggregate	1/2 weeks		1/month per active plant
Mix Air Content	1/lane mile at placement	1/2 weeks at placement	As needed to control operations	1/2 weeks	2/lane mile at placement	1/month at placement

*Acceptance testing is to be performed at the process control frequency until uniform material is produced.
**Test for mix control purposes, not for acceptance.

TABLE OF MATERIALS TESTING AND FREQUENCY

	Without Contractor Process Control		With Contractor Process Control			
Portland Cement Concrete Pavement	**Acceptance Testing**	**Independent Assurance Testing**	**Process Control Testing**	**Verification/ Check Testing**	**Acceptance Testing ***	**Independent Assurance Testing**
Mix Slump	1/lane mile at placement	1/2 weeks	As needed to control operations	1/2 weeks	2/day at placement	1/month at placement
28 Day Compressive Strength	1 set/lane mile at placement	1/2 weeks	As needed to control operations	1/2 weeks	1/lane mile at placement	1/month at placement
28 Day Tensile Strength	1 set/lane mile at placement	1/2 weeks	As needed to control operations	1/2 weeks	1/lane mile	1/month at placement

*Acceptance testing is to be performed at the process control frequency until uniform material is produced.

TABLE OF MATERIALS TESTING AND FREQUENCY

	Without Contractor Process Control		With Contractor Process Control			
Portland Cement Concrete Pavement	**Acceptance Testing**	**Independent Assurance Testing**	**Process Control Testing**	**Verification/ Check Testing**	**Acceptance Testing ***	**Independent Assurance Testing**
Mix Design Gradation						
Coarse Aggregate	1/100 CY	1/2 weeks	1 per 1/2 day production	1/2 weeks	1/500 CY	1/month
Fine Aggregate	1/100 CY	1/2 weeks	1/100 CY	1/2 weeks	1/500 CY	1/month
Moisture Content**	1/ 1/2 day/stockpile aggregate		1 per 1/2 day/ stockpile			
Mix Air Content	1/25 CY	1/2 weeks	As required to control operation		1/25 CY	1/2 weeks

*Acceptance testing is to be performed at the process control frequency until uniform material is produced.
**Test for mix control purposes, not for acceptance.

TABLE OF MATERIALS TESTING AND FREQUENCY

	Without Contractor Process Control		With Contractor Process Control			
Portland Cement Concrete Pavement	**Acceptance Testing**	**Independent Assurance Testing**	**Process Control Testing**	**Verification/ Check Testing**	**Acceptance Testing ***	**Independent Assurance Testing**
Mix Slump	1/25 CY	1/2 weeks	As required to control operation		1/25 CY	1/2 weeks
28 Day Compressive Strength	1 set /100 CY (minimum 1 set/ pour)	1/2 weeks	As required to control operation	1/2 weeks	1/set/100CY minimum 1 set/pour)	1/2 weeks

*Acceptance testing is to be performed at the process control frequency until uniform material is produced.

TABLE OF MATERIALS TESTING AND FREQUENCY

			Supplier Process Control			
Hydraulic Cements and Additives			**Process Control Testing**	**Verification/ Check Testing**		
Quality of Portland Cement Or Lime Or Pozzolans				1 per lot	Agency wide review—1 per supplier per month	
Quality of Air Entrainer Retarder Water Reducer Accelerator Super Plasticizer			1 per lot	Agency wide review—1 per supplier per month		

TABLE OF MATERIALS TESTING AND FREQUENCY

			Supplier Process Control			
Reinforcing Steel and Wire Rope			**Process Control Testing**	**Verification/ Check Testing**		
Physical and Chemical Properties						
Billet, Rail, and Axle, Steel Bar			1/heat number	3 bars/size/ 10 tons		
OR						
Fabricated Steel or Rod Mats						
OR						
Dowel and Tie Bars						

TABLE OF MATERIALS TESTING AND FREQUENCY

			Supplier Process Control			
Reinforcing Steel and Wire Rope			**Process Control Testing**	**Verification/ Check Testing**		
Physical Properties						
Welded Steel Wire Fabric			1/heat number	1/75,000 SF		
Epoxy Coated Steel Bars			1/lot	3 bars/size/lot		
Wire Rope or Wire Cable or Rod Mats			1/heat number	1/shipment		

TABLE OF MATERIALS TESTING AND FREQUENCY

	Without Supplier Process Control		With Supplier Process Control			
Concrete and Metal Pipe	**Acceptance Testing**	**Independent Assurance Testing**	**Process Control Testing**	**Verification/ Check Testing**	**Acceptance Testing ***	**Independent Assurance Testing**
Physical Properties Quality Concrete Pipe	1/supplier/lot		1 per lot	Agency wide—1 per supplier per month	1/supplier per lot	
Chemical and Physical Properties Quality Metal Pipe	1/supplier/lot		1 per lot	Agency wide—1 per supplier per month		

*Acceptance testing is to be performed at the process control frequency until uniform material is produced.

TABLE OF MATERIALS TESTING AND FREQUENCY

	Without Contractor Process Control		With Contractor Process Control			
Miscellaneous Material/Test	**Acceptance Testing**	**Independent Assurance Testing**	**Process Control Testing**	**Verification/ Check Testing**	**Acceptance Testing ***	**Independent Assurance Testing**
Quality of Material	1/supplier/lot		As required to control operation		Agency wide 1/supplier/ month	

*Acceptance testing is to be performed at the process control frequency until uniform material is produced.